I0487719

PHYSICS DOESN'T NEED TO BE *THAT* DIFFICULT

PHYSICS DOESN'T NEED TO BE *THAT* DIFFICULT

A Proposal for

A NEW PHYSICS MODEL

J.F. Bennett

Copyright © 2009 by J.F. Bennett.

Library of Congress Control Number: 2009908911
ISBN: Hardcover   978-1-4415-6952-3
                Softcover   978-1-4415-6951-6

All rights reserved. No part of this book may be reproduced or transmitted in any form or by any means, electronic or mechanical, including photocopying, recording, or by any information storage and retrieval system, without permission in writing from the copyright owner.

This book was printed in the United States of America.

To order additional copies of this book, contact:
Xlibris Corporation
1-888-795-4274
www.Xlibris.com
Orders@Xlibris.com
68104

CONTENTS

1

Why this Book was Written

The physicists deemed newsworthy these days are often those whose focus is far out. Astronomical subjects like the Big Bang, the Expanding Universe, Black Holes, and Dark Matter are all the rage. Yet I have developed a conviction that what physics most needs right now is a new look at some things closer to home: how students are taught the subject and the conventional model on which that teaching is now based. I came to that conviction in an unexpected way.

When I retired at age 65 from a full-time career in finance, including service in the US Treasury as Under Secretary for Monetary Affairs and at Exxon Corporation for fourteen years as chief financial officer and director, I became a director of a number of other corporations, including the US computer company Tandem and the European electronics company Philips. It occurred to me then that some up-to-date knowledge of physics might come in handy during my service on the boards of those companies. It had been almost fifty years since my brief brush with formal education in the subject, an undergraduate semester course at Yale. So I began reading some textbooks and other books available from public libraries and book stores. It turned out that during my period of service no questions ever came before the boards where a better knowledge of physics seemed required, but by the time I realized that I had become interested in the subject.

Unfortunately I had no access to university libraries, but I did have an advantage relative to university students when I came to a statement I couldn't understand. I had the time—and the long airline flights—to continue to wrestle with the statement rather than have to memorize it for the next exam and pass on to the next subject. And I had the good fortune to find some knowledgeable experts who had great patience with my resistance to education and who were willing in occasional meetings to try to explain to me the conventional answers to questions I asked. One of these tutors was a brilliant MIT post-graduate student, Derin Sherman, later a professor at Cornell College in Iowa. Another was Howard Hayden, a perceptive professor of physics at the University of Connecticut, now retired to Colorado. A fourth was Jill Moring, an experienced bio-chemist. And the fifth was a thoughtful professor from Hampshire College, Herbert Bernstein. These scholars did not share the extreme skepticism which I developed about current theory and instructional practices, but all of them were tolerant, and Professor Hayden had edited an alternative academic journal, *Galilean Electrodynamics*, had published a number of scientific journal articles questioning the conventional interpretation of some empirical data as supporting Einstein's Special Relativity.

Despite the professional help I received I found trying to understand the way the theory of physics is being presented to students today extremely difficult. My experience convinced me that the subject could be made much simpler and more understandable. Sometimes the complex formulations of the experts reminded me of those frequent historical occasions when religious leaders objected to presentation of the teachings of their sects in a form which the uninitiated might understand without priestly intermediaries. Often I found myself wondering whether the emperor had any clothes.

The bit of history just related explains why I started studying physics at an advanced curmudgeonly age but doesn't explain why I stuck with the subject long after the original rationale was no longer relevant and why I have written this document. There were several reasons. For one thing it occurred to me that something in writing might make it seem less queer to my friends and relations why I had spent so many hours looking perplexed. (Now sometimes when I am just daydreaming my family leaves me undisturbed since they assume I'm pondering some deep physics mystery.) Secondly, I found out way back, when I was a teaching fellow in money and banking at Harvard, that I learned more from trying to expound a subject to others than when I simply studied it. I also hope, when I write down what others believe to be ridiculous, that some kind expert will take the time to explain to me the error of my ways. In Washington I became accustomed to the process of making policy by passing around drafts of a proposed speech for clearance. Yet the most important reason for writing this paper was my conviction, which I wanted to try to convince some others to share, that the teaching of physics is in serious need of revision.

I did not begin my belated study of physics in the expectation that I would end up questioning many of the pedagogical procedures now in use, and I certainly had no expectation that I would eventually conclude that much of the model on which that teaching is based should be replaced. But eventually I came to have considerable sympathy for the advice given in a guest editorial in the American Journal of Physics in February 2009 by Professor Craig F. Bohren that anyone working on an introductory textbook should "adopt the working hypothesis that everything in previous textbooks is wrong." I stand in awe before what has been accomplished over the centuries by careful observers, measurers, and experimenters in enlightening the rest of us about the nature of our physical world. Yet I am not only underwhelmed by the theoretical model which the experts have produced to summarize the observed data but I also feel there are serious shortcomings in the way the subject is taught.

Much of the text available to me has been, not about current theory, but about history, particularly about who did and who didn't win a Nobel Prize in physics. Although many of the writers of that text seemed to be suffering from acute Nobel envy, I am not saying that the history of physics is not instructive and interesting. What I do say is that modern physics theory is so complex and inter-related that it should be taught first without distracting the students with anecdotes about the personal idiosyncrasies, family problems, and religious beliefs of past physicists and with accounts of all the travails and false starts encountered over the centuries in the development of that theory. Study of the history should come later. With a better understanding of the current physics model a student could gain more from the history. That was the sequence used in the attempt to teach me graduate economics, and I found it superior. But in the field of physics I could never find a concise statement of the theory.

I also came soon to the conclusion that it is unfortunate that the teaching of physics has become so exclusively organized around what the physicists call energy. The beginning student must first learn to live with the unfortunate fact that the profession is using a definition of work and energy different from that used by most of us in commerce and in daily life. Unlike the physicist everybody else thinks he has used more energy when, with a constant straining for two minutes, he has pushed a car 25 feet than when he pushed a motorbike the same distance with the same straining for 30 seconds. For the physicist the force times the distance, and therefore by his definition the work done and energy applied, were the same in both cases if the friction resistance were negligible. For the physicist what the straining accomplished in both cases was the same magnitude of the product .5MV², which is called kinetic energy. The bike with its lower mass acquired more velocity so that the product of half that magnitude and the squared velocity for the bike was the same as the same product for the car with its greater mass and smaller squared final velocity. Yet that measure of what force had accomplished is not easy to get the mind around. M for mass is easy. V for velocity is easy. And their product MV for momentum is not hard. But it is not easy for the student to acquire the same intuitive comprehension of .5MV². Unfortunately when a mass has a force applied to it for a time the product of that force and the time has been given the name impulse, a word which for some people carries with it implications of brevity and irrationality. Yet it is my impression that education in physics would be aided by much more frequent use of the term impulse.

Another problem is that physicists use the word energy in various senses. Sometimes when a physicist refers to energy he is referring to the influence emanating from a charged source; at other times he is referring to the joint effect on some charged target body of an incoming influence and influence coming from that body. Similarly physicists sometimes use the word force to refer to the influence coming from a charged source, sometimes to refer to a joint effect on a target body, and sometimes to a small body which has been ejected from a larger body.

In order to speak the physicist's language the student must also learn to consider that no work is being done when, for example, he holds up one end of a sofa for a minute so that his mother can sweep underneath. For the physicist the sofa was moving no distance during that minute, so no work was being done. But the student would have been inclined to consider that he had expended energy as he applied a force for a period of time even though that force was just offset by the downward force of gravity on the sofa.

In addition to simplicity, as compared to the physicist's energy, another advantage of momentum is that it has a direction which must be specified. The physicists do recognize—though rarely mention—that the kinetic energy of a body is relative, since they recognize that a ball thrown forward out of a moving car window has more energy relative to a standing pedestrian than relative to a car ahead moving in the same direction. Kinetic energy must be measured relative to some reference frame. But physicists treat kinetic energy as something which doesn't have a direction.

Another complication is that, although rarely made clear, measurements of the momentum and of the energy of a body are usually, not only using different definitions, but also measuring quite different things. It would be possible, though it is rarely done, to calculate the momentum of each of a composite body's components in their various directions and then simply to add them up without any offsetting. But momentum is usually calculated just for the total mass of a body in one direction. On the other hand, it would be possible to calculate the energy of the total mass of a body in just one direction, but energy is usually considered to be the sum of the energies of the components of the body in various directions without any offsetting for movements on opposite directions. As a result when, for example, a dumbbell is sitting and spinning on a table with its center not moving with respect to the table the physicists say that the dumbbell has total energy which includes the separate energies of the dumbbell balls moving in opposite directions and of each of the moving electrons, quarks, etc. in the dumbbells. And if the balls are spun faster or the particle components of the dumbbell are made to move faster by heating, then the movements in opposite directions are not considered to offset each other and the body is said to have acquired more total energy.

When two charged bodies are in a position to influence each other they affect each other's velocities, but after any amount of time the combined momentum of the two bodies remains unchanged. The increase in the momentum of one is just offset by the decrease of the momentum of the other. But after any time period in which the ability of the bodies to affect each has not been effectively exhausted the total kinetic energy of the bodies is not maintained. The physicists have in this situation invented a concept of potential energy which allows them to say that total energy in a closed system is always maintained when potential energy is taken into account. The potential energy of a system at a particular time is just the difference between the total kinetic energy at that time and what the total kinetic energy would be if the bodies of the system were allowed to influence each other until something happened which prevented any further influences from passing between the bodies. As a result a system of bodies can be calculated to have various different potential energies depending on what future happening is assumed.

To make matters even more difficult for the student the physicists decided to consider that a system composed of two like-charged bodies pushing each other apart without limit have zero potential energy. Then as the bodies pick up speed their increasing kinetic energy is said to be just offset by progressively more negative amounts of potential energy.

Adding even further to the confusion the texts say that, in addition to kinetic energy from the movement, if any, of a body as a whole with respect to its frame of reference, in addition to kinetic energy which the body might have as a result of movement of its component parts, and in addition to any potential energy, the body also has some mysterious attribute known as rest energy. The dumbbell mentioned above would have had this rest energy even if the dumbbell were not moving with respect to the table on which it were sitting, the dumbbell's component parts had been brought to a stop, and there were no forces at play between components of the dumbbell. This rest energy was defined as mc^2, where m is the mass of the body and c is said to be the velocity of light. In this formulation rest energy and the mass of the body were related by a constant, so that an increase in mass meant an increase in energy. With this relationship it was not much of a stretch to say that rest energy and mass were the same thing. To me, however, it seems unnecessary and undesirable to subvert a good concept and good word, energy, by treating a body which is not moving either as a whole, or even in part, as having energy. As indicated in the discussion of the new model, the need for mysterious rest energy does not arise with what I regard as probably more reasonable and more accurate hypotheses relating to the behavior and effects of electrical influence.

Some new textbooks no longer say that a large body's mass changes when its velocity changes. That convention was never necessary and was pedagogically disastrous. But that old usage still persists in discussion of nuclear reactions. When, for example, a neutron breaks apart into a proton, an electron, and a neutrino it is said that the total mass of the daughter products would be less than the mass of the parent neutron unless the mass of the daughter products is considered to increase because of their increased velocity after the split-up. In this situation I am led to wonder, au contraire, whether some of the proton combination's former mass actually flew away in photons with mass. However, in many texts' discussions there is often no mention at all of what things really are flying out of interactions in which energy is said to be radiated. Energy is treated almost as if it were some sort of newly created liquid which is squirted out. And even when photons are mentioned as the conveyors of energy there seems to be no realization of the difficulty for the student when he is told the photons involved sometime appear out of nowhere and convey energy even though it is said that they would have no mass at rest.

There are some circumstances in which knowledge of the calculated energy of a body may be of more use than knowledge of its momentum. When, for example, a propeller is lowered into a container of liquid knowledge of the kinetic energy arising from the propeller's revolution may help predict how much the temperature of the liquid will be increased, but I am convinced that in most cases when a force accelerates a body consideration of the momentum given the body would be more beneficial than discussion of the energy given to the body.

I began to have a collection of words which, like energy, as opposed to kinetic energy, seemed to lead to confusion when used in the texts. Among these words, for example, were space warp, annihilation, mediation, flux, and decay. With respect to the first of those words, I doubt that I am the only one who can't comprehend how nothing can be warped. Yet for those of us not fully indoctrinated the word space is another word for nothing. I realize that the experts say that when photons from a star pass by the sun their velocity remains constant but that their path is altered by a warping of space caused by the great mass of the sun. Wouldn't it be simpler—and just as accurate—just to say that the path of moving photons can be affected in some circumstances by influences emanating from the sun and that these forces can affect the path of an approaching photon, and would also increase its speed were it not for the slowing effect of increasing collisions with the molecules of gas clouds as the photon gets nearer to the sun?

I was also bothered by many other conventional model conclusions that were in conflict with ordinary observations in our everyday "macro" world. For example, would there really be attraction between two like-charged particles moving side-by-side when those same two particles would repel each other when at rest with respect to each other? And would a positive-charged body really be accelerated away to a lesser extent when approaching another positively-charged body than when the two bodies were at rest with respect to each other the same distance apart? Many physics experts seem to feel that such surprising counter-intuitive mysteries in their version of the "micro" world give their profession a distinction. I recognize that their conclusions could be correct, but I still give significant weight to observations of the behavior of larger objects when I reflect on the imperfections of the observations supporting contradictory conclusions for the "micro" world.

In view of such thoughts I gradually came to the conclusion that the appropriate course for me was not just to build up a list of complaints but to try to develop an outline of what would seem to me to be a more understandable alternative presentation of the theory of physics. This I started to do and ultimately ended up with the document which follows, **A New Physics Model.** That section does not present any new empirical evidence. It does attempt to suggest how existing data could be presented in a more helpful form. And it does reject, as will be specified, a number of basic assumptions of the conventional presentations. The description of the new model represents the type of presentation which I wish I had had available when I began my belated study of the subject. The new model should, of course, not be accepted to the extent it is shown to be contradicted by empirical evidence.

The suggested new model does accept many of the features of the conventional model, such, for example, as the conclusion that time independent of a specified clock is meaningless. And the new model accepts the conventional reading of most important empirical observations.

Early on in trying to develop the new model I came to the conclusion, however, that the physical universe with which we have contact would be a lot easier to understand if there were considered to be, apart from neutrinos and far out dark matter, only two basic sub-atomic building blocks, that all other things we encounter were considered to be combinations of those two small particles, and that there were not many different types of action-at-a-distance force in our world but only Coulomb influences arising in various combinations from those two types of small particle. That approach differs from the current practice of treating quarks, electrons, positrons, and photons as well as neutrinos and some evanescent other things as basic particles. If a proton was composed of a properly layered distribution of positive and negative particles that structure by itself could explain why the influence emanating from a proton and a neutron appears to be short-range without the necessity of assuming the existence of a peculiar strong force entirely separate from the electric force. And the fact that photons, which have no net charge, sometimes attach themselves to electrons would be easier to understand if a photon were considered to be a combination of an electron and a positron and thus to have a negative charge on one part of its exterior and a positive charge on another part of its exterior.

Some physicists now include in their thinking virtual photons, strings, and membranes the existence of which have never been observed. I happen to think those constructs are not useful, but if they were I would see no objection to including them in a model even though they have never been observed.

It has seemed to me that a model which makes use of a composite photon with mass can help put an end to another intellectual burden now put upon the student, that of having to consider simultaneously that light is particles, that is photons, and is also a wave in some substance-less material. As is discussed at some length in the model section of this document, if a photon is a combination of a positive and a negative part, then when the photon revolves the projected oscillating electric influences can be considered to be light while the movement of photons themselves can bring the source of that light to new locations. That arrangement would be easy for the student to comprehend. Combinations of electrons and photons could also help to explain "excited" states of electrons orbiting nuclei and the increase in mass which bodies sometime experience. I realized that the conventional model has somehow reached the conclusion that photons do not have any mass, and yet photons are the result of the combination of two particles which do have mass In a collision a photon can cause movement in another body which does have mass. The addition of photons to a body has been observed to increase both the temperature and the mass of the body.

Another area that seems to me to cry out for a new treatment is the teaching of magnetism and electricity. Today teaching in that area starts off with a confusing presentation of two different patterns of what are called field lines. For a central charged body surrounded by some smaller bodies of opposite charge the lines from the outer bodies point inward like the spokes from the rim to the hub of a wagon wheel indicating the direction in which the central charge would pull the outer bodies. Lines for a magnet indicating the direction in which a central magnet would pull some surrounding magnets would look exactly the same if the outer magnets were free to turn. Yet that fact is not mentioned, and the texts switch without adequate explanation to a new usage which presents the magnet's lines, not as lines pointing in the directions in which a second magnet would be pulled, but rather the as lines indicating the direction toward which each second magnet's north end would point if it were mounted on and free only to turn on an axis perpendicular to a line between the magnets. The texts also usually say, incorrectly, that the string-like positions taken up by iron filings sprinkled on a page around a magnet are representations of the field lines from the magnet. In fact the position of each filing is the net result of the influences on that filing not only from the main magnet but also from the temporary magnetism of all the many other filings on the page.

Many of the texts do say that electricity and magnetism are two aspects of a common force but I have yet to come across, either in textbooks or in books on electricity and magnetism, a discussion which provides an understandable elaboration of that statement. Sometimes I have come across statements that magnetism is the result of electrons spinning and orbiting near each other in the same direction, but never have I seen presented an explanation how two negative particles could attract each other. Also I have heard statements that magnetism can only be explained by use of Relativity, but again I have never seen an understandable explanation of those statements.

In relation to electricity and magnetism the texts treat the Scots physicist James Clerk Maxwell with veneration of an intensity usually reserved for a saint. He it was who established the practice of considering that a moving body with an electric charge created around it two different influences, one magnetic and one electric. Yet I am convinced that the subject would be significantly more understandable if a particular particle, whether moving or not, were considered to radiate only one type of influence, which could influence charged bodies and magnets differently depending on their nature and motion.

Current texts follow a practice sanctified by Maxwell of referring to a field surrounding a charged body or a magnet. I find that to be confusing since the impression is given of something real and stable in the space involved whereas in reality each influence from a charged body or a magnet is in rapid motion away from the body. And I find it easier to work with a model in which those influences in time have an effect at a distance than to consider that fields are something real, something more than a set of reference numbers.

Another burden which the student must bear today is the treatment of quantum physics, which is made to appear very mysterious. Some writers on physics have been dining out with breathless accounts of how difficult it all is. Yet it seems to me that the quantum aspects of current physics need not be presented in so mysterious a fashion. It is merely necessary to take cognizance of the apparent fact that electrons, positrons, and photons come only in certain sizes. For example, an electron can be understood to move out to a particular different orbiting level when a photon is attached and adds a certain amount of mass to the combination without reference to any mysterious constant or to mysterious permitted levels. And the various sets of confusingly denominated shells and sub-shells and the so-called quantum numbers are pedagogically disastrous. Without them it would be easier to comprehend that different numbers of photons can become attached to an orbiting electron/photon combination, that two bodies of substance can't be at the same place at the same time, that the components of an atom or molecule can affect each other, and that it makes a difference whether an orbiting combination is, or is not, revolving around its nucleus in the same direction as its photons are circling about its orbiting electron.

Sometimes the texts now even go so far as to state that something doesn't happen in nature because of some man-made rule rather than to state that such-and-such man-made rule seems to reflect what is actually happening in nature.

The texts also make heavy weather out of the fact that it is not now possible to find out the specific location and velocity of some particular very small combination. But what is so mysterious about that fact or the fact that, by reasoning from the observed behavior of a large number of small things, it is possible to come up with an estimate of the probability that a thing of that type will be a certain distance from a nucleus?

Eventually, in trying to develop a more understandable model I came to the heretical conclusion that Einstein's Relativity should be dislodged from its central role in the current teaching model and that some of the hypotheses of Relativity should not be included in the new model at all. I realize that in the view of most physicists that conclusion automatically turns me into a fit subject for abuse as a "crackpot". Fortunately avoiding that label is not a prerequisite for my continuing to receive my pensions. I came to a realization that Relativity is not necessary to explain why it becomes increasingly difficult to accelerate a body further as its velocity increases, why wires carrying currents in the same direction attract each other, and why magnets have the effects they do. When I combined that fact with a realization of the seemingly unreasonable nature and inadequate empirical support of some relativity hypotheses I concluded that they should not be included in the proposed model. I realize that in omitting Relativity I have also rejected the hypothesis of the sainted Maxwell and deified Einstein that all light and other electric radiation should always be measured on various bodies as traveling at the same speed regardless of the relative movement of the various bodies.

In view of the major role the Special Relativity hypotheses play today in the conventional model I added an extensive separate section at the end of the presentation of the proposed new model to illustrate in greater detail the hypotheses and some of the implications of Special Relativity. Unfortunately, in view of the nature of Special Relativity, the illustration is necessarily complex!

A model without all of Relativity can be far simpler than the currently conventional model. There would, for example, no longer be any need for rest energy. But that is not the only reason why I rejected much of Relativity. Rather I acted as a result of a combination of simple judgmental doubt about some of the consequences of the Relativity hypotheses and awareness of the imperfections—perhaps even flimsiness—of the support provided by the empirical observations usually referenced in support of Relativity. I should make clear, however, that my rejection of Relativity does not, in my view, conflict with my acceptance of the observations that the path of photons can be affected by Coulomb force, that a man in a box couldn't tell whether his feet were being pressed against the floor of a supported box by gravity pulling him down or by acceleration of the box being pushed up by some force, and that Coulomb influences and photons generally travel at about the same high velocity.

One of the Relativity consequences which offends my simple judgment is that acceleration of a body should affect clocks at different locations on the body to different degrees as seen by observers on an un-accelerated body, and, so far as I am aware, no one has provided convincing evidence of having observed this result. And yet this result is necessary if all bodies are to measure incoming flashes of light to have the same velocity.

Another questionable but logical consequence of Relativity is that, again as seen by observers on an un-accelerated body, an accelerated body should continue to appear to have its length compressed even after the acceleration has ended. This consequence has apparently also never been observed.

According to Relativity, as noted earlier, a positively-charged body receiving influence from another positively-charged body will be accelerated less when the bodies are approaching each other than when they are at rest with respect to each other. This seems upside down to me though I realize that with the high velocities involved this is a matter difficult to test by planned experiment.

The fact that clocks can be slowed when their shells are accelerated, or their shells are held in place against greater gravitational pull, is not an issue, but my judgment recoils from the Relativity conclusion that such clocks will continue to run slow once the acceleration has ended. Here again I don't think this conclusion has ever been directly observed experimentally with man-made clocks, but two types of observation have been interpreted as support for the Relativity treatment of clocks. As one type it has been contended, that the treatment is validated by observations comparing the half-lives of two types of evanescent combinations, muons and pions, descending to the earth from the stratosphere with half-lives of the same types of particles created in a laboratory. The muons, for example, have been observed to have a half-life of 2.2×10^{-6} seconds when created on earth but have been judged to have a half-life about nine times longer when they are created by incoming cosmic radiation reportedly high in the atmosphere and travel at high speed down to the surface of the earth. This fact has been interpreted to mean that the muons have the equivalent of internal clocks which in the case of the muons coming down to the earth appear to have been slowed by acceleration and continued at their slower pace, so that on those "clocks" the half-life was really the same as the half-life of muons created nearer the earth. Yet I find it easy to suspect that there were features of the experience of these muons which would not be common with the experience of all accelerated clocks.

The other type of observation adduced in support of the continuing slow advance of a clock which has been accelerated was first arranged in the Hafele-Keating 1971 experiment in which four cesium clocks were synchronized on the ground and then at separate times flown in commercial airliners eastward and westward around the world. The changes in times on these clocks during their flights were then compared to changes in clocks at the Naval Observatory in Washington. Attempts to interpret the results of the experiment were complicated by several factors. There were differences among the clocks even when flying in the same direction. The flying clocks were not coasting in continuous orbits. The clocks spent various periods of time on the ground at airports. The clocks took different amounts of time to circle the earth, about 65 hours for those traveling eastward and about 80 hours for those traveling westward. The clocks flew at different velocities relative to the earth. (Presumably a factor which contributed to the longer time for the westward trip was that the planes carrying the clocks flew mostly in latitudes where the prevailing wind was from the West). And, of course, the flying clocks were held up on average at a higher altitude than the clocks left on the ground. The eastbound and westbound clocks actually flew at different altitudes, and latitudes, while in the air.

It was realized that the time-keeping of clocks held up at a higher altitude advances faster, and an attempt was made to adjust the observed data for this fact. (No attempt was made to adjust for the different latitudes for the planes.) After the altitude adjustment the eastbound clocks were calculated to have slowed by 59 nanoseconds relative to the ground clocks and the westbound clocks to have gone faster than the ground clocks by 273 nanoseconds. Relative to an observer standing next to the ground clocks this data could not be considered as support for the Relativity hypothesis of apparent slowing and continued slower motion of accelerated clocks since, even though all the flying clocks were accelerated relative to that observer and traveled long distances, some of the flying clocks advanced rather than slowed relative to the ground clocks.

Some have argued that the slowing and speed-up of the clocks on the planes should not be calculated relative to the earth clocks, which should not be considered to be in a stationary system of coordinates since those clocks were accelerating as the earth spun on its axis. (This comment would seem to cast doubt on all experiments involving clocks on the earth's surface.) These supporters of Relativity contended that the pace of the clocks should be compared to a theoretical clock located at the center of the earth—or at least somewhere along the line about which the earth spins—and that, since the earth spins eastward, the eastbound flying clocks would have a higher velocity relative to that theoretical clock than relative to the earth while the westbound flying clocks, whose eastward spin was to some extent offset by their flying westward, would have a lower velocity relative to the theoretical clock. These commentators argue that when the data is considered relative to a theoretical clock at the center of the earth the experiment supports Relativity. Yet a clock at the center of the earth would be moving in its orbit about the sun. Under the circumstances I see the Hafele-Keating results, not a support for Special Relativity, but rather as a denial of General Relativity.

Particularly controversial in the proposed new model is my assumption that the speed of Coulomb influences approaching a body is influenced by the relative speeds of that body and the source of those influences. Most texts consider that subject was settled by observations, notably by Willem DeSitter, that light rays from two rotating binary stars moving in opposite directions appear to reach the earth at the same speed. I have read some rare articles by physicists questioning the conclusions drawn from those observations, but my own contrarian view derives from a combination of "common sense" reaction, suspicion that incoming light rays from the stars were pushed toward uniformity by collisions with other small bodies on their very long journey through space and the earth's atmosphere, and awareness of the difficulty of moving sources to sufficiently high speed to test the subject in earthly experiments. And, as explained in a separate section of this booklet, in adopting in the proposed model's treatment of the velocity of light I took comfort from the fact that the famous Michelson/

Morley experiment did not rule out the possibility that the velocity of reflected light is affected by the velocity of the reflecting surface.

In having the temerity to disagree with the current consensus among physicists by proposing a new and simpler physics model, especially one which rejects some of the teachings of Newton, Maxwell, and Einstein, I have taken some comfort from the knowledge that most leading physicists once believed in a luminiferous ether. I am also aware of the comment of the famous physics professor Richard Feynman that "There is always another way to say the same thing that doesn't look at all like the way you said it before. I don't know what the reason for this is. I think it is somehow a representation of the simplicity of nature." And I have been encouraged by the remark of the Harvard philosopher/physicist Percy Bridgman: "Whatever may be one's opinion as to the simplicity of either the laws or the material structures of Nature there can be no question that the possessors of such conviction have a real advantage in the race for physical discovery."

2

A New Physics Model

A model of the basic materials of our physical universe, and the interactions among them, can assist in predictions of the consequences of events we observe or instigate. Such a model should attempt to summarize the results of past observations into a comprehensible form while being consistent with those observations. As experience with an existing model and new data become available it may be appropriate to revise or replace that model. The author is convinced that the model which is now in general current use by teachers of physics is in need of replacement and could be replaced by a far simpler yet more comprehensive model. Presented below is an attempt to present an outline of such a model suitable for the year 2009. The presentation is not intended as a textbook. That would require more detail, more examples, and pretty pictures. But the author believes the presentation below would be well suited for anyone who wished to gain a broad understanding of what is now known about basic aspects of our world.

Space, Time, and Motion

The space in which we live appears to contain myriads of mostly-moving material things between which there is nothing material, but somehow influences can travel through this space from one thing to another. It has never been possible, however, to isolate a portion of space in a container in which there was a perfect vacuum in which there were absolutely no things or moving influences.[1] (The preceding superscript number and similar numbers at the ends of some subsequent paragraphs refer to numbered paragraphs in the following section titled **Footnotes on Differences in the Proposed Model from the Conventional Model.**)

No one of the things in space, nor any combination of them, has a legitimate claim to be considered the reference against which the magnitude, position and movement of other things should always be measured. The magnitude, position and movement of one thing can be measured relative to any other thing by use of multiples or fractions of the distances and directions among points considered to be at rest with respect to each other on any thing or things chosen, for convenience, as the measures of distance and direction.

There is no specially ordained standard for measurement of the time duration of any movement or for the time interval between two events. A measurement of the time duration of the movement, or lack of movement, of some thing relative to another thing is just a comparison of the behavior of those two things relative to the behavior of two other material thing chosen as reference. For scientific purposes the time taken, for example for a ball to fall from the top of a tower to the ground, is usually measured by comparison of the movement of the ball to the earth and the movement of some other selected thing such as the arm on a clock as that arm moves between two positions on the clock face when the velocity of the movement of the arm is believed to be acceptably uniform. There is no standard, however, by which any movement such as the movement of a clock arm can be proved to be exactly uniform. For scientific purposes time duration has no meaning apart from the physical processes by which it is measured, but when a number of different processes show the same repetitive motion within the degree of precision required that motion is considered to be an acceptable measure of time duration. The pace of advance of a clock may be changed, however, for example, by friction when the case of the clock is pushed or pulled against the inner mechanism of the clock, but clocks usually resume their earlier pace of advance when the force which caused the slowing is ended. When events take place at two appreciably different places those events can be considered simultaneous only by some assumption as to the velocities with which information had been transmitted from the two places to a single place or as to what happened when a timepiece was transported among the places.

The direction of time is by established usage considered to be the direction of some natural events such as movement of a mature apple from a tree branch to the ground or the exit of a chick from an egg.

The magnitude of the movement of a thing relative the movement in a timepiece can be estimated by humans, and when a human observer says that two events which took place very close to him clearly did not occur at the same time his statement of which event came first in time may be accepted. But if two events, even though taking place close to an observer, did not clearly take place at different times his judgment as to which came first or that the events were simultaneous may not be accepted. A person's judgment of the time duration between events will rarely be accepted even if the two events occurred at about the same place. [2]

When one body moves a certain straight-line distance in a certain direction with respect to a second thing in a certain measured time interval and then moves that same distance again in the same direction with respect to the same thing in a time interval of the same size the first body is said to have a constant average velocity, and not to be accelerating, with respect to the second thing. That velocity may be zero, and the body or the other thing may or may not be moving with respect to other things. If the body had moved a different distance in the same direction with respect to the same thing in an equal second time period the body would be said to have had acceleration in that direction with respect to the reference thing. In physics, the term acceleration can refer to an increase in velocity, to a decrease in velocity, that is a deceleration, and to a change in direction. In some circumstances a body may be said to be accelerating even when the body seems not to be moving. For example, when a ball is tossed straight up in the air it appears not to be moving at the highest point reached, but at that time the motion of the ball is changing from up to down along the vertical direction. When one body is accelerating with respect to a second body the second body is also, by definition, accelerating with respect to the first body, but, it is sometime important to know which body had been most recently affected by an influence incoming from some third body.

When a body moves in a non-linear fashion with respect to a second body then the second body necessarily moves in a non-linear fashion with respect to the first body. When a body rotates with respect to a second body then, even if the distance between the mid-points of the bodies doesn't change and even if the second body had felt no force, relative to the first body the second body can be said to move in a circle around the first body

The velocity of a body and the direction of its motion are not magnitudes which are directly measured. Rather they are calculated from a body's measured positions at two different times. These measurements are necessarily imprecise. As a result statements of a body's average velocity and direction tend to be more accurate the further apart the positions being measured and the further apart the times.

When a body moves in one direction in a time period, say 10 degrees east of north relative to the earth's surface and then moves in a different direction in the next time period, say 25 degrees east of north, it is not possible to express the body's acceleration and direction by single numbers relative to the earth. It would be possible, however, to calculate the body's acceleration in the northward direction and its acceleration in the eastward direction relative to the earth

The words velocity and acceleration are also used with different meanings in one special circumstance. When something is orbiting a central body the line from the central body to that thing swings through an angle over time relative to any stationary straight line through the middle of the central body in the same plane. The rate of growth in that angle from some chosen base line is known as angular velocity. The rate of growth in that angle in one period over the growth of the previous period is known as angular acceleration over those periods.

Particles, Charge, Influence, Force, and Mass

Given the present state of our knowledge it is useful to work with a model which considers that, at least in our solar system, all the things which we consider material, other than possibly some extremely minute things called neutrinos to be discussed later, are now either one or the other of two types of substance or combinations of the two types. In this booklet an agglomeration of either type of substance without intermixture with any of the other type will be called a particle, and the word particle will not be used to refer to any other type of body. One type is called an electron and it is said to have a negative charge. The other type is called a positron and it is said to have a positive charge. Apart possibly from the neutrino, no body, other than a particle, composed of just one type of substance has ever been observed. Yet a combination body can be composed of a larger number of one type of particle than of the other type and thus can have a net negative or net positive nature.[3]

The strength of the negative attribute of each electron appears to be the same. The positive strength of each positron appears to be the same. And the strengths of an electron and a positron appear to be of the same magnitude though of different nature. There is, however, an important difference between an electron and a positron: electrons have been observed not combined to another particle or combination of particles for a long time but no positron has ever been observed to have a long separate existence. Observed unattached positrons have always quickly combined with some positive particle or particles. It has never been possible to create a new electron or positron or destroy or break apart an existing one.

When two particles of the same type are at rest sufficiently close to one another, not touching each other and not appreciably close to an unequal concentration of other particles, and an effective shield between the particles is removed then the two particles move apart after a time interval and at an initial acceleration the magnitudes of both of which depend on how far apart the particles were. The time delay and the magnitude of the initial acceleration are, within the degree of precision of measurement now available, the same whether the two particles are both electrons or both positrons. The time delay is proportional to the distance between the particles. When the two particles are of opposite types they begin after the same time delay to accelerate toward each other at the same initial rate as two like particles accelerate apart. When one of the two particles is replaced by two of the same type of particles side-by-side then the other single particle is accelerated twice as fast. If the two particles at rest side-by-side are of opposite sign then the particles are not accelerated. When the distance between two particles at rest is twice as large then the particles are accelerated one quarter as much. In other words acceleration in this situation varies inversely with the square of the distance between the particles.

In the inter-actions just described the model assumes that some influences passed directly between the particles although no evidence has ever been observed that the influences were conveyed by movement of any material bodies between the particles. (It is possible, however, as discussed at length below, for a particle or combination of particles to be broken off or fly off from a larger combination of particles and then for the separated particle or combination of particles to be a separate source of positive and/or negative influence.)[4]

Various words are used in describing the direct inter-actions between particles. The influence which emanates from the charge of a particle and inter-acts with charges on other particles is sometimes referred to as electro-magnetic. At other times it is called Coulomb influence. This influence may be measured not only for the influence emanating from a single particle but also for the net influence from a combination of particles. The charge on 1.6×10^{19} particles of the same type is said to be one Coulomb. It is said that each electron has a negative charge of -1.6×10^{-19} Coulombs and each positron has a positive charge of 1.6×10^{-19} Coulombs.

Since two particles of opposite type which have been at rest with respect to each other don't accelerate at an infinitely rapid rate when a barrier between them is removed, but each does accelerate at the same rate, it is said that each particle has the same resistance to being moved, i.e. the same inertia, which is said to result from each particle having the same mass. The proposed model considers to be material any thing which contains mass and charge, neither of which can be destroyed, and neither of which ever appears without the other. For a body which is a combination of particles the mass is proportional to the number of its charges. Its net charge depends on the relative number of the two different types of charge. A particle's mass is not a feature of the particle unrelated to the charges of its components. Rather the mass of a particle is just a word which indicates the way in which the total charge of that particle affects its reaction over time to the relationship between its net charge and the net charge of another nearby body. The charge and the mass of a particle do not change over time. And removing a single particle from a combination inevitably both reduces the mass of the combination and changes the net charge of the combination. The mass of a single particle is said to be a certain fraction of the total mass of the particles of a metal cylinder defined by international agreement as having a mass of one kilogram (kg). On this basis both the electron and positron are each said to have a mass of approximately 9.11×10^{-31} kilograms (kg).

Combinations of particles may be broken apart but there has never been an observation of a single particle being broken apart or destroyed, and the strength of the influence emanating from a particle does not appear to diminish or vary over time. When there are no obstructions influence is simultaneously projected from a particle in all directions. The influence is projected by a particle whether it is at rest, moving at a steady pace, or accelerating.

When Coulomb influence moving in a certain direction reaches the center of the mass of a body with a net charge, if the effect of that influence is not offset by the effect of other influences, then the body will be given instantaneous acceleration and, if the influence continues to be applied, over time the velocity of the body will increase and the distance the body has traveled in that direction will increase.

The word force is used to convey the combined effect on a particular body of influences emanating from the charges of that particular body plus the influences from the charges of other bodies whose influences are reaching the particular body. Force cannot be measured, however, without reference to mass. When a force is doubled the instantaneous acceleration of a body with a given mass is doubled, and when the mass of a body is doubled a given force will provide only half as much acceleration.

When two bodies are stationary with respect to each other and one body is experiencing a Coulomb influence from the other the relationship between the force on the body receiving the influence, the mass M of that body, and its instantaneous initial acceleration A can be expressed by the simple equation:

$$F = M\,A$$

Alternatively when high speeds are not involved the relationship between force and mass can be approximated by considering what happens when a constant force is applied for a specified short time T to a given mass. The result is that the mass is given a specific increase ΔV in its velocity, as expressed in the following equation produced by multiplying both sides of the previous equation by T:

$$F\,T = M\,A\,T = M\,\Delta V$$

since with a constant force and low speeds the increase in velocity over a time period is approximately the product of the instantaneous rate of acceleration and the time period.

A third way to define and measure force and mass, again when high speeds are not involved, is by calculating what happens when a single constant force has caused a body to travel a certain distance in the direction of the force. This can be accomplished by recognizing that when a body is pushed by a constant force over a time the increase in the average velocity of the body over that time is approximately half the final increase in velocity and the increase in the distance traveled is the product of the average velocity increase and the time, as expressed in the following equations, in which both sides of the previous equation have been multiplied by the average increase in velocity:

$$F\,T\,(\Delta V/2) = M\,\Delta V\,(\Delta V/2)$$

$$F\,D = .5\,M\,\Delta V^2$$

Coulomb influence always travels in a vacuum in relation to its source particle at the same velocity, approximately 300,000 kilometers per second, a velocity known as c, whether the source is stationary or moving with respect to other bodies. When influence from a particle has traveled to another body then as measured on that body the velocity with which the influence traveled toward it is the sum of c and the velocity, if any, with which the source particle and the target body moved toward each other. The velocity with which influences travel between bodies may be different from the velocity of a charged combination which has been broken away from the source body. [5]

When a body emits Coulomb influence that influence does not cause any recoil of the emitting body. The velocity of an influence and its effect on another body are not affected by any movement of the source after the time the influence is projected. The effect of a particular influence portion is not affected by any inter-actions which may be occurring simultaneously as a result of other portions of influence which have moved in other directions from the same source at the same time. Once an influence from one particle has an effect on another particle the incoming influence has no further effects.

As will be discussed in more detail in a later section, the things which we are aware of encountering in daily life are usually combinations of both types of particles. No combination composed solely of particles of one type has ever been observed. The combinations encountered in practice are composed of both types of particles held near each other or touching each other by the "glue" of the attraction of the oppositely-charged particles. A combination can, however, have unequal quantities of the two types of particles and empty spaces between some of the constituent particles, so that the combination can have a net overall charge greater than one. The mass of a combination is the sum of the masses of the constituent particles, but the effective net charge of a combination, when it is not so near to another combination that the positions of the particles within the combination is significant, is close to the net of the component positive and negative charges when one type of charge is considered a positive number and the other type is considered a negative number. When inter-acting combinations are close together, however, the effect of their charges may be affected by the shape of the distribution of the particles within the combinations. It has been noted that a combination with a spherical shape with its particles distributed uniformly on its exterior inter-acts as if all the combination's net charge were concentrated at its center. The interactions of individual combinations when extremely close together can differ significantly, however, from those described above.

Despite what is known about particles there is much which is not known and not specified in the model. Nothing is known about what particles are made of, what their shape is, what gives rise to the projected Coulomb influences, how the influences from a particle cause repulsion and attraction of other particles, why influences travel at a velocity of c with respect to their sources, why the amount of influence projected by a particle does not weaken over time, why the left side of a particle does not repel away the possibly like-charged right side, and why particles frequently combine into certain combinations and not into others.

Apart from the effects of the two particles which have been described above there have been some observations of effects seemingly caused by a collision with a larger combination of a very small thing to which the name neutrino has been given. Its size and mass would be extremely small relative to the size of an electron or a positron though it has never been possible to determine that size and mass with any precision. No positive or negative characteristics have ever been associated with the effects of a neutrino. Conceivably it is a combination of a small amount of positive substance and a small equal amount of negative substance, or perhaps it is a separate type of basic substance. There are indications that millions of them are passing through our bodies each second without apparent effect, like buckshot through chicken wire only more so. Study of neutrinos is at a very early stage and neutrinos today play a very limited role in explaining physical events.

The Effects of the Movement of Particles

When Body One and Body Two are at rest with respect to each other and Body Two is being affected by influence incoming from Body One, the effect on Body Two depends both on the types and magnitude of the charges in the two bodies but also on how much of the influence from Body One "hits" particles in Body Two. Since Body One's influence travels in all directions that influence is shared at any particular distance from the source by the area of the surface of a sphere, an area which varies with four pi times the square of the distance from the center of the sphere. The share of Body One's total influence which "hits" Body Two, depends therefore on both on the square of its distance from Body One and the area of Body Two perpendicular to the direction of the force. The strength of influence emanating from a particle does not diminish as that influence moves away from its source but the influence is spread out and the proportion of that influence affecting a target whose charges are spread over an area of a particular size varies inversely with the square of the distance from the source to the target.

When two bodies with net charges have been at rest with respect to each other some distance apart and an effective shielding between the bodies is removed the calculated initial rate of acceleration of, say, body two when influence from body one arrives is only instantaneous. That acceleration can not be simply multiplied by some amount of time to calculate precisely the velocity which body two would achieve after that amount of time. If, for example, the bodies both had net positive charges the acceleration of body two would immediately start to decline from its initial rate for two reasons. First, the distance between the bodies would be increasing as they accelerated apart so that later influences had further to travel and spread out and as a result those influences would be weaker. And, second, the movement of the bodies would result in later incoming Coulomb influence "hitting" body two with lesser velocity relative to body two, causing less acceleration in body two.

When two bodies which both have the same type of charge are at rest with respect to each other, and body two is receiving Coulomb influence from body one, the initial acceleration of body two is larger the larger is the charge of body one. And the acceleration of body two is larger the larger is the charge of body two. The proposed model assumes that the acceleration of body two is proportional to the product of the two charges. The model also assumes that the acceleration of body two is inversely proportional to its mass. And, since the influence from body one is spread out over time, the model assumes that the acceleration of body two is also inversely proportional to the square of the time taken for the influence to travel between the bodies. These considerations can be presented in an equation:

$$A_I = Q_I \times Q_{II} \times K / (M_{II} \times T^2)$$

In this equation the K represents a constant which can be estimated empirically and depends on the units of measurement used for the different magnitudes in the equation. In this case if the separation of the bodies is assumed to be r, and, since the model assumes that Coulomb influence travels with a velocity of c with respect to its source, the time dimension equation can be restated as shown in the revised equation:

$$A_I = Q_I \times Q_{II} \times K / (M_{II} \times [r/c]^2)$$

In this particular case if it is assumed that the charges, the mass and the constant all happened to have values of 1, that r had a magnitude of 6 meters(m), and that c has a value of .3 meters per nanosecond(/m/ns), then the initial acceleration of body two would be:

$$A_I = 1 \times 1 \times 1 / (1 \times [6/.3]^2) = .0025 m/ns/ns$$

If, however, in the body were moving to the right with a velocity V of .1 m/ns relative to the source of Coulomb influence on body one the acceleration of body two can be calculated with an equation similar to that used above for initially stationary bodies but two changes have to be made. These changes are signaled by the change in the new equation for the symbols for the constant, from K to k, and for the time from T to t;

$$A_I = Q_I \times Q_{II} \times k / (M_{II} \times t^2)$$

The value of the constant has to change because in this second situation the Coulomb influence would "hit" body two with lesser velocity and therefore lesser consequence. The model assumes that the consequence is proportional to the velocity with which the Coulomb influence reaches its target. The ratio of k to K would then be the ratio of the Coulomb velocities with respect to body two in the two cases:

$$k / K = (c - V) / c$$

$$k = (1 - V/c) K$$

On the basis of that equivalence the acceleration equation can be restated as;

$$A_I = Q_I \times Q_{II} \times (1 - V/c) \times K / (M_{II} \times t^2)$$

In this second case the value of the time t can be calculated as:

$$t = [r + t \times V] /c$$

$$t = r /[c - V]$$

In this case the acceleration equation becomes:

$$A_I = Q_I \times Q_{II} \times (1 - V/c) \times K / (M_{II} \times \{ [r / [c - V])^2$$

With the same assumptions as in the earlier case about magnitudes the acceleration equation becomes:

$$A_1 = 1 \times 1 \times (1 - .1/.3) \times 1 / (1 \times \{ [6 / [.3 - .1])^2 = .00074$$

The acceleration would thus be less when the Coulomb force had further to travel and arrived at lesser velocity relative to its target.

If the movement of the bodies had been toward each other, rather than away, the acceleration of the right body would have been greater than in Case I. When a target body is moving perpendicularly to the direction of the incoming influence the magnitude of the acceleration of the body is not affected.

The second numerical example above illustrates why it becomes increasingly difficult to use a single source of Coulomb influence repeatedly to accelerate a like-charged target away and up to a very high velocity and impossible in this way to produce in a target a velocity in excess of c relative to a stationary source. There is, however, no maximum speed limit. It would still be possible, although it has never been demonstrated in practice, to produce a measured speed to the right in a charged target body in excess of c relative to the position of a like-charged source to the left of the target when an influence was emitted if the source had first been accelerated to a considerable velocity to the right relative to a reference frame on which the positions and motions of the bodies were measured. And if two combinations were each moving toward or away from each other at a velocity only slightly in excess of one half of c relative to some reference frame the pieces would be moving relative to each other at a velocity in excess of c as measured on that reference frame.[6]

The factors illustrated in the examples have an important role, as outlined in a later section of this paper, in explaining the behavior of magnets and wires carrying electric currents.

As explained above, the accelerating effect on a charged body of influence originating from a particular source is affected over time both by changes in the distance the influence has to travel and by the velocity, if any, of the body receiving the influence relative to the velocity in the same direction of the incoming influence. In everyday life, however, it is sometimes possible to make useful calculations without taking into account either changes in the distance the influence involved has to travel or the relative velocity of the source and target of the influence. When, for example, the distance changes are small, the velocity of a body is small relative to the velocity of an incoming Coulomb influence, and the time T the force is applied is small then the resulting change in velocity ΔV can in practice be approximated simply:

$$F \times T = M \times A \times T = M \times \Delta V$$

And in the special circumstance in which the velocity of the target body was zero at the beginning of the time period involved the increase in the target's velocity and its final velocity V are the same, so that:

$$F \times T = M \times V$$

When, for example, an apple falls from a tree branch the source of the gravitational influence attracting the apple can be considered to be the center of the earth, and that is so far away that the difference between the distance from the branch to the center of the earth and the distance from the ground under the branch to the center of the earth can for most purposes be ignored. And the velocity of the apple is so small relative to the velocity of the gravitational influence that the effect of the apple's velocity on its acceleration can be ignored.

In that case the average velocity of the body over the time period of the apple's fall is one half the final velocity and the distance traveled by the body while the influence was being applied is equal to the product of the time period and one half of the final velocity:

$$D = T \times V/2$$

A charged body will not accelerate even though subject to Coulomb influences if those influences are applied in opposite directions to the same point and fully counter act each other. A person standing on a platform will not accelerate with respect to the platform if he is being pulled toward the platform by the influence of gravity but is being pushed equally strongly in the other direction by influences from within the platform when his weight seeks to bend a portion of the platform downward. The feet of a person standing in an elevator will not accelerate relative to the floor of the elevator even though both the person and the floor may be moving at a constant velocity with respect to the building housing the elevator. When, however, an elevator is accelerated upward by being pulled by a cable that accelerating force is not applied directly to all parts of a person in the elevator. While the bottom of his feet won't move with respect to the elevator floor his upper body may be compressed downward a bit while the accelerating force is transmitted upward from his feet through his body.

A person who was in a compartment without windows and was feeling pressure between his feet and the floor could not tell whether he was being pulled down by gravity while the box were being held up or he was out in space where there was no appreciable net gravitational influence and something was pushing the box toward his feet. If he were in such a compartment and felt no pressure in any direction he could surmise that there was no non-gravity pressure on the compartment but he could not tell whether he was out in space where there was no appreciable net gravitational influence or he and the box were being equally accelerated in some direction by a gravitational influence affecting both him and the box equally.

When influence applied to a body is not directed toward its center of mass that body will experience angular acceleration, that is an increase in its rate of spin about its center of mass, and will also accelerate at the same time in the direction of the influence. The name torque has been given to the product of the magnitude of a force on a body and the shortest distance from the point about which the body is being spun by the force to the line of movement of the force. When a temporary influence has given spin and torque to a body then the body will have acquired what is called angular momentum. The rate of spin and the torque will then remain constant until some other influence acts upon the body. For a given influence the torque is greater the greater the perpendicular distance from the path of the influence to the point about which the body is spinning. Rotation does not occur when a body experiences pushes in opposite direction by forces each of which would by itself create the same amounts of torque in opposite directions even though the pushes are not of the same magnitude. For example, if a force of 2 applied 5 feet from the center of mass of a bar pushed it in a clockwise direction with a torque of 10 then a force of only 1 applied 10 feet from the center in a counter-clockwise direction would apply a countervailing torque of 10 and leave the bar at rest.

When a charged body, for example a positive one, is on a course to move past a second positively-charged body without a collision then as the bodies come closer together the second body is given simultaneously a push in the direction of the motion of the first body and a push away from the path of the first body. Later as the distance between the bodies increases the second body is pushed simultaneously back in the direction from which the first body came and further away from the path of the first body. But the net effect is to move the second body somewhat in the direction of the first body's motion, as well as away. This happens because the influence pushing the second body ahead is enhanced while the bodies are growing nearer, while the effect of the influence pushing the second body backward is reduced while the bodies are moving apart.

When a Coulomb influence is applied to only one of two bodies and the relationship between the bodies is made to change it sometimes makes a difference which body received the incoming influence. This fact is illustrated by the two cases shown on the **Diagram 1**, which plots the position of two positively-charged bodies at two different times relative to the reference frame represented by the page.

Diagram 1

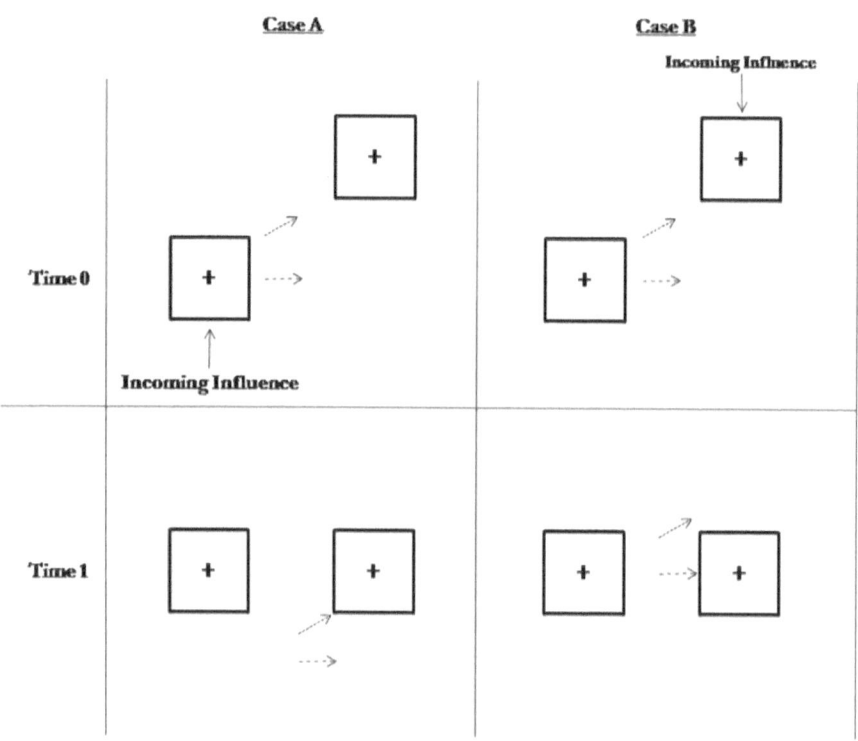

At Time 0 in both cases the bodies are at rest with respect to each other and have the same relationship with respect to each other but in Case A an influence is applied at that time to push the left body upwards while in Case B an influence pushes the right body down. For the left bodies two arrows, one solid and one with dashed lines, are shown pointing to the right to indicate at Time 0 and Time 1 the direction of motion of two selected positive Coulomb influences emanating from those bodies at Time 0. By

Time 1 when the influences from the left arrive at the right the bodies have in both cases the same position side-by-side and the same motion relative to each other, but the right bodies are then about to be pushed by the Coulomb influences in different directions in the two cases relative to the reference body. In Case A the right body is about to be pushed up and to the right relative to the reference frame, while in Case B the right body is about to be pushed directly away to the right relative to the reference body.

When force is applied to the center of mass of a moving body in a direction other than that in which the body is moving the resulting direction and motion of the body will then be composites of the original direction and motion and the direction and motion of the force.

When a constant attractive influence from a source is applied continuously to a body perpendicular to the body's direction of motion the path of the body will in some circumstances be converted into an orbit about the source of the influence. It is possible, for example, for a satellite without a motor to orbit the earth continuously in any chosen direction at any chosen altitude if the satellite has been made to move perpendicularly relative to the line from the satellite to the center to the earth at a velocity appropriate to that altitude and direction. In this case the gravitational influences reaching the satellite from the earth at any point are just enough to bend the path of the satellite from a straight line tangent to the earth at that point over to a circular path around the earth. Influences from the sun and other heavenly bodies make no appreciable combination to this result. The sun can, however, provide a convenient reference frame for analysis. And, in view of the fact that, relative to the sun, the side of the earth facing the sun spins roughly eastward about an axis running between the earth's north and south poles, if there were above a point on the equator a satellite which had been given an appropriately-sized eastward push at just the right altitude, 35,786 km., then that satellite could remain directly above the chosen point on the equator at a constant altitude. Many communication satellites have been put into what are called geo-centric orbits about the earth on this basis.

Similarly it can be noted that a point facing the sun on the surface of the earth in the northern hemisphere moves relative to the sun in an eastward circular direction about the earth's polar axis. A point on the earth's surface north of the first point but on the same line of longitude also moves in a circle but, being closer to the axis, moves with lesser velocity. If then a long-range cannon at the first point fired a shell directly north the path of that shell would be affected both by its northward propulsion and by its eastward motion. Since that eastward motion of the shell would be greater than that of the northern point, which would be closer to the earth's north/south axis, if the shell fell back to the ground at the latitude of the northern point the shell would then be to the east of the northern point. This is a consideration which long-range gunners have long known to take into account.

Bodies may be accelerated relative to each other even if there is no significant influence passing between the bodies. What is necessary for this to happen is that the composites of the external influences affecting the two bodies, whatever the sources of those influences, are not proportional to their masses and headed in the same direction.

If the charges in a combination are not uniformly distributed then an incoming Coulomb influence from a combination with the same type of net charge may push that combination as a whole in a direction other than directly away from the incoming influence, may even pull the combination toward the source of the incoming influence, and may cause the combination to spin.

A combination with a net charge will experience alternating net attraction and repulsion if it is near to a spinning combination which has its axis of spin perpendicular to the line between the combinations and has a distribution of positive and negative components on opposite sides of that axis. The spinning contribution could, for example, be a simple combination of one electron and one positron. Even though the alternating type of influence might be discernible at the non-spinning combination the rate of spin might be so rapid that no net forward and backward acceleration of the non-spinning combination would be discernible. An outgoing alternation or oscillation of repulsive and attractive influences can also result when some area is provided first with a majority of positive charge and then with a majority of negative charge. As will be discussed later, such alternations have many important practical uses.

When a sizeable Coulomb influence is projected against a combination the components of that combination may be pushed or pulled into a new configuration inside that combination. For example, an incoming net positive influence could pull the negative components in a combination toward the source of the influence and push the positive components away. Similarly when an incoming influence is projected against a collection of combinations with various net charges the relative positions of these combinations with respect to each other can be changed. And when a series of oscillating influences is experienced by a combination the components in that combination may be induced to have spins which match those of the combination which gave rise to the incoming oscillating influences. When that happens radiation of influences from the induced spins may in turn have effects on yet other combinations.

Impulse, Momentum, Work, and Energy

The product of the magnitude of a force and the time of its application is called impulse by physicists. A force applied over a time provides an impulse whether or not the body to which the force is applied is moved since a body receiving equal and directly opposite impulses will not move. But, if the source of the incoming influence is a force and the target of the force are not moving rapidly with respect to each other when an impulse arising from a constant force affects a body not subject to any offsetting influences the body's velocity is increased approximately in accordance with the relationship mentioned earlier:

$$F \times T = M \times \Delta V$$

The product of the magnitude of a mass and its velocity is called momentum. When therefore a body is receiving only one constant force, as in the equation above, the impulse and the increase in momentum are equal. In the special situation where a body had no velocity at the beginning of a period and the force was constant during the period the impulse would equal the body's final momentum:

$$F \times T = M \times V$$

Physicists use the term work for the product of a constant force and the distance over which it is applied. As indicated by the equation explained earlier, if there are no other offsetting forces involved, when such a force is applied it's magnitude is equal to one half the product of the mass involved and the square of the increase in the final velocity of the mass:

$$F \times D = .5 \times M \times \Delta V^2$$

and in the special situation where a body had no velocity at the beginning of a period and the force was constant during the period the work would equal one half the product of the mass and the body's final momentum:

$$F \times D = .5 \times M \times V^2$$

In this case the physicists say that work of $.5 \times M \times V^2$ has been done to move the particle the distance involved and to increase its velocity from zero to its final value. The $.5 \times M \times V^2$ is said to be the kinetic energy of the particle at the end of the period. In kinetic energy thus defined velocity has enhanced importance relative to its role in momentum since velocity is squared in the definition of kinetic energy.

If two bodies, of different masses, composed of particles rigidly tied together were initially at rest with respect to one another and after a period ended up moving in the same direction with the same kinetic energy then the bodies would have felt different forces and would then have different velocities at the end of the period. When each of two stationary rigid combinations with different masses has the same force applied to it until it has traversed the same distance in the same direction as the other the resulting product of force and distance is the same for both bodies so the work applied to each is said by the physicists to be the same. $F \times D$ and $.5 \times M \times V^2$ are the same for both. This physicist's definition of work done differs from that most commonly used. A homeowner usually thinks he has done more work when he has overcome a car's inertia and pushed it fifteen feet up to a certain velocity with constant straining than when he has with the same straining pushed a motorcycle the same distance up to a higher velocity in a much shorter time. The physicist's measure of work is the same whether the mass of a body moved a certain distance by a given force is large or small. The physicist thinks in terms of force applied for a certain distance, and if the body receiving a force didn't move there was no work done. For the physicist if a person pushes a thing and the opposing force of friction prevents the thing from moving then no work was done. Most others think of work in terms of effort expended over time whatever that effort achieved.

A difference between momentum and kinetic energy for a body which has been put into motion from a stationary state is that the former is normally calculated by physicists as a product of the total mass of the body and its movement in some prescribed direction even though it would be theoretically possible to calculate the momentums of the different parts of the body and add up those separate momentums algebraically for a total for the body. Theoretically the total kinetic energy for a body is calculated separately with respect to each component of the body, and the separate component energies are simply added up without any offsetting for movements in opposite directions. Kinetic energy is thus considered to be a scalar quantity rather than a vector quantity. In practice it is never possible to measure the total energy of a composite body with precision since it is never possible to learn the precise motions of all the small components in various directions. While kinetic energy is not a vector quantity kinetic energy calculations must be made with respect to some reference frame. Yet, even though the forces involved in calculating a body's kinetic energy have directions, physicists do not consider that a body's total kinetic energy has a direction. For physicists there is no such thing as a body with negative kinetic energy. If there were two bodies moving in opposite directions the total of their kinetic energy would be simply the sum of their two separate positive energies. When a body is moving with kinetic energy in one direction a force applied to the body in the other direction will reduce the body's energy in the original direction. If the force continues long enough to do more than reduce in the energy in the original direction to zero the body then acquires not negative energy but positive energy in the other direction.

When a body is subject to a constant force over a period of time the average rate at which impetus is given to the body is equal to the force times the increase in momentum divided by the time:

$$F \times T/T = F = M \times \Delta V/T$$

When work has been done on a body, and its kinetic energy increased, by a constant force the average rate of increase of the kinetic energy over a time is known as power:

$$.5 \times M \times \Delta V^2/T$$

When charged Combination One projects influence toward charged Combination Two and charged Combination Two simultaneously projects influence toward Combination One it is likely, but not inevitable. that both bodies will receive a force. At some point near Combination One, for example, a barrier might be interposed after the influence from Combination One had passed and before the influence from Combination Two had arrived. In that case Combination One would not recoil when it emitted influence toward Combination Two and would not receive influence from Combination Two. But when there are no barriers in between the combinations and charged Combination One projects influence toward charged Combination Two then at the same time Combination Two projects influence toward Combination One and then each body will receive influence. If the combinations are initially stationary with respect to each other and the component parts of the combinations have permanently fixed positions within the combinations, when the influences originate and arrive—always assuming neither influence is blocked en route—the influences arriving at each combination are of the same absolute magnitude since both are the products of the same charges and are affected by the same distance of separation. If both combinations also have the same mass each mass will experience the same sequence of accelerations—in opposite directions—relative to a reference body. If instead Combination One has twice the mass of Combination Two then the increase in the velocity of Combination One relative to a reference body will be half the increase in the velocity of Combination Two relative to that reference body. But the product of mass and induced velocity is the same for both combinations. The change in momentum—in opposite directions—has the same magnitude for both bodies regardless of the relative sizes of the masses involved.

If velocity in one direction is given a positive sign and velocity in the opposite direction is given a negative sign then when two like-charged combinations which have been at rest with respect to each other are influencing each other one combination will be given a positive change in its momentum and the other will be given a negative change in momentum of the same absolute magnitude relative to a reference body. It is sometimes important to note the change in the momentum of each separate body but it is also sometimes important to note that the algebraic sum of the changes in momentums for a system consisting of two interacting bodies will be zero. That statement can be generalized into the rule that the algebraic total momentum, that is the net momentum, of such a system relative to a reference body cannot be changed by forces internal to the system whether or not the combinations are moving with respect to each other and with respect to the reference body. A system's total algebraic momentum relative to a reference body can be changed only by forces originating outside the system. When a flying artillery shell explodes its component pieces move in different directions but the total momentum of all the pieces does not change relative to the earth.

The momentum of a system not affected by external forces is maintained in all directions. For example, if positive Combination A, which was moving toward the East past stationary positive combination B somewhat to the south, projected repulsive force toward B, while B projected a repulsive force toward A, and afterwards A moved toward the Northeast while B moved toward the Southeast, then the algebraic sum of the momentum components of the two combinations in the East direction after A passed by B would be the same as before the collision, and the algebraic sum of the components in the North/South directions would be zero after the combinations passed by each other just as that sum was before.

When two combinations with like net charges are moving directly toward each other their mutual repulsion causes both to slow down. In most cases they will slow to a halt with respect to each other and then start moving apart without touching. Since, as will be discussed later, even when combinations without net charges are involved most combinations have on their surfaces net negative charges which repel each other strongly as the combinations come very close together. Many apparent collisions are like the one described above and do not involve actual physical contact between the particles involved, even though that may not be the impression we have when watching a collision with our non-microscopic eyes. Even when combinations do touch, however, and temporarily compress each other they may not become entangled with each other and quickly start moving apart. As bodies move apart at each distance of separation, when the bodies are not moving at a speed approaching c, the forces of repulsion are approximately the same when the combinations are going apart as when they are coming together. Not surprisingly therefore, it has been observed that when two combinations come very close together without becoming entangled with each other they then move apart with approximately the same velocities with respect to each other as they came together while the sum of the momentums of the two combinations remains the same.

When two bodies come together and stick together the two bodies move as one after the collision, and the momentum of the combined body after the collision is the same as the algebraic sum of the separate momentums before the collision.

If a positively-charged combination directly approaches a second less-massive positively-charged combination without becoming entangled then as a result of the influences passing between the bodies they will separate after the collision with approximately the same sequence of absolute differences in their velocities as they had while coming together. In this case the more-massive combination will continue to move in its original direction but at a reduced velocity, and the less-massive combination will move in the same direction at a greater velocity than it previously had. If the approaching combination is less-massive the combinations will still move apart after the collision at approximately the same relative speeds with which they came together, but in this case the incoming combination will have its velocity reduced and possibly reversed.

Some of the relationships described in the preceding paragraphs can be illustrated by a simple example. Suppose there were two positively charged rigid bodies: one on the left with a mass of 15 grams and one on the right with a mass of 5 grams. Initially the bodies were very far apart. The one on the left was moving to the right with a velocity of 5 meters per second and the one on the right was moving to the right with a velocity of 2 meters per second along the same line of motion as the body on the left. At that time the momentum of each body and of the sum of the two are shown below:

$$M_L \ \times \ V_L \ + \ M_R \ \times \ V_R \ = \ \text{Combined Momentum}$$
$$15g \ \times \ 5m/s \ + \ 5g \ \times \ 2m/s \ = \ 85 \ gm/s$$

In this case when the bodies were so far apart the Coulomb influences passing between them were having no measurable effect, but gradually as the bodies got closer together each one began to be appreciably influenced by the influence from the other. Since at each moment each body was the same distance from the other, and the product of the charges determining the strength of the force on each body was the same, each body experienced a sequence of forces of the same magnitude. The body on the left experienced a force which reduced its velocity to the right. The body on the right experienced a force which increased its velocity to the right. Since the body on the right had a smaller mass its change in velocity was greater, but, since the changes in velocity were inversely proportional to the masses, the magnitude of the change in momentum was the same for both bodies but of opposite sign. Eventually the bodies reached their closest approach to each other when the slowing of the left body and the hastening of the right body brought their velocities to the same level. At that time the ratio of the accumulated change in the velocity of the left body to the accumulated change in the velocity of right body was the inverse of the ratio of their masses. Their momentums were then:

$$15g \ \times \ (5 - .75)m/s \ + 5g \ \times \ (2 + 2.25)m/s \ = \ \text{Combined Momentum}$$
$$15 \ \times \ 4.25m/s \ + 5g \ \times \ 4.25m/s \ = \ \text{Combined Momentum}$$
$$63.75gm/s \ + \ 21.25gm/s \ = \ 85gm/s$$

The combined momentum of the bodies remained unchanged.

The bodies would then continue to the right while pulling apart. Eventually they would be apart the same distance as originally and the forces on them would no longer be large enough to cause any further noticeable changes in their velocities. At that time the separate momentums would be:

15g x (5 - .75 - .75)m/s + 5g x (2 + 2.25 + 2.25)m/s = Combined
                                                                                                                     Momentums
15g x   3.5 m/s   + 5g x   6.50m/s   = Combined
                                                                                                                     Momentums
        52.5 gm/s   +   32.5gm/s   = 85gm/s

and the combined momentums of the bodies would have remained the same. The sequence of forces acting on each body after their closest coming together would have been the same as the forces prior to their closest coming together but in reverse order.

The results of particles coming together would be exactly like those described in the previous paragraph, however, only when the bodies were rigid, so that none of the Coulomb forces involved caused some parts of the bodies to move differently than other parts, only when the velocities were not great, and only when the events occurred in a vacuum. When a body moves in air some of its momentum is lost to the extent the body pushes air molecules in the direction the body was moving, thus transferring momentum to the air molecules. But the extent to which air molecules are pushed sideways out of a body's path does not reduce the body's forward momentum since the air molecules on the right push the body to the left to the same extent the air molecules on the left push the body to the right.

The total kinetic energy of the two rigid bodies in the example above was at any time the sum of the kinetic energies of the two bodies at that time. If—unrealistically—the components of each body were tied rigidly together and always moved in exactly the same direction as each other initially the kinetic energy was:

$$.5 \text{ x } 15g \text{ x } 5^2 \text{ m}^2/\text{s}^2 + .5 \text{ x } 5g \text{ x } 2^2 \text{ m}^2/\text{s}^2 = 197.5$$

The kinetic energy of the bodies was not, however, the same at all later times. At the closest point of the bodies, for example, their kinetic energy was:

$$.5 \text{ x } 15g \text{ x } 4.25^2 \text{ m}^2/\text{s}^2 + .5 \text{ x } 5g \text{ x } 4.25^2 \text{ m}^2/\text{s}^2 = 180.625 \text{ m}^2/\text{s}^2$$

This example illustrates the fact that when charged particles or bodies interact with one another their total kinetic energy is not usually maintained at the same level during the period of interaction. In the example, however, if the bodies were allowed to push each other apart until they returned to their positions at the start of their interaction, when there would no longer be any significant interaction, then the total kinetic energy of the bodies would have risen back to its initial amount. It could be said that when the bodies were at their closest approach they had kinetic energy of 180.625 m^2/s^2 and the potential of adding another 16.875 m^2/s^2 if the bodies were allowed to continue influencing each other until they returned to their initial positions.

In one special case the interaction of charged bodies can lead to increasing total kinetic energy for a while and then extinction of that kinetic energy. When, for example, two oppositely—charged stationary bodies of equal mass are far apart and attracting each other but haven't yet begun perceptible motion toward each other the bodies have effectively zero kinetic energy. As the bodies then gradually pick up speed in moving toward each other their total kinetic energy is increased. But when they reach each other if they then merge with one another the merged bodies would again have no velocity and no kinetic energy.

The amount of force needed to initiate a change a body's position enough so that the body can pick up a large amount of kinetic energy may be very small in relation to the magnitude of the gain in kinetic energy. This would happen, for example, when a small force pushed a cannonball off the flat top of a high slender column so that the ball could then be freely influenced by gravity as it fell to the ground. It would also happen, as often seen in chemical and nuclear reactions, when bodies at rest with respect to each other at certain distances of separation accelerate each other when the separation is made only slightly different in either direction by the effect of some slight external force. For example, suppose that body A, as shown below, had a positive charge of ten while rigid body B had a negative charge of one at its left end and a positive charge of ten at its right end, as pictured below:

```
    A                        B
(+10)        (-1)                        (+10)
    I------------.4625 --- I----------------1-----------------I
```

These two bodies would remain at rest with respect to each other if the distance between A and B's left charge just happened to be, as indicated above, .4625 distance units and the distance between A and B's right charge happened to be 1.4625 distance units even though B and A both had net positive charges. With these distances the attraction between A and the left end of B would just offset the repulsion between A and the right end of B, leaving the bodies at rest with respect to each other. But if B were given the slightest push to the right, increasing the distance between the two bodies they would then repel each other and B's velocity to the right would continue to increase the further it moved from A. On the other hand if B were given the slightest push to the left the bodies would then attract each other and for a while B would pick up velocity and energy to the left. In this situation when the bodies were at rest with respect to each other it would not be possible to say that B had just one potential for increasing its kinetic energy.

In practice the component parts of a combination are usually not rigidly fixed in place within the combination, and these parts may be differentially affected by incoming Coulomb influence, but if these parts all remain within the combination their plus and minus induced momentums will offset one another and have no effect on the total momentum resulting from the overall motion of the combination. But any action which stirred up the component parts of a combination without knocking any of those parts away would increase the kinetic energy of those parts and thus the total kinetic energy of the combination.[7]

Combinations of Particles

Photons

The smallest known combination of particles is composed of one positron and one electron held together by the effects of their opposite charges. Such combinations are known as photons. They have a mass of approximately 18.22×10^{-31} kg equal to the sum of the mass of an electron and the mass of a positron. Photons are usually extremely stable over time, though they are sometimes made to break apart into their constituent positrons and electrons. Photons are usually rotating about an internal axis. The axes of rotation can point in any direction, and the observed rates of rotation vary over a very wide range.

A photon has no net electric charge, and photons rarely become attached to one another, though two photons can become attached if they are spinning at the same pace in a manner which usually puts the positive charge of one photon near the negative charge of the other and the negative charge of the former photon usually near the positive charge of the other. And a photon can sometimes become attached to a larger charged body.

When a photon is reflected from or broken away from a body the photon proceeds in a straight line relative to that body unless the photon while traveling is affected by forces from or a collision with another body.

As will be discussed later, photons reflected from or emitted from larger bodies are the principal source of the oscillating electric influences which create visible light and are also a major source of other forms of oscillating electric radiation.

Sometimes a photon encountered by an eye or other detector on the earth had been created when an electron and a positron already on the earth came together and became attached. In some cases a photon encountered on the earth came from a distant star. But in most cases photons met on the earth originated in the sun and then traveled to the earth. Sometimes such photons are encountered as they arrive for the first time on the earth. In most cases, however, such photons have already been reflected on one or more occasions from a surface on the earth or have broken off from an earthly body to which they had in some way been attached earlier.

Apparently in most, if not all cases, photons leave the sun with a velocity close to c, that is 300,000 km/sec, relative to the sun, perhaps as a result of the nature of nuclear reactions within the sun which gave rise to the photons. Since the distance from the earth to the sun changes very slowly relative to the velocity of photons coming to the earth those photons seem to arrive at the earth with a velocity of approximately c also with respect to the earth.

A photon can be reflected from another body either as a result of an actual collision or as a result of a temporary orbiting of the body. When a spinning photon collides head-on with another body the photon is reflected directly back away from the body with approximately the same velocity relative to the body as the photon had when approaching the body. When a photon does not collide with another body but the photon comes close enough, and has an appropriate rate of spin, to be "captured" by the body, an electron, for example, the photon can go into orbit about the body at least for a time. In this case the photon could continue to have its incoming velocity relative to any reference frame as the photon orbits the body. And the photon could retain its velocity if it simply escaped from its "captive" orbit. When a photon collides with a body or goes into orbit around it the photon probably transfers some of its momentum to the body, thereby reducing the photon's going away velocity below the arriving velocity, but in practice the mass of a photon is usually so very small compared to the mass of the other body that any resulting change in the photon's velocity as a result of transfer of momentum is not discernible.

A point on the earth's surface on the side toward the sun has in its orbital motion around the sun a westward movement relative to the sun. This westward movement is greater than the point's eastward movement as a result of the earth's spinning relative to the sun. When a photon arriving at the earth with a velocity of c with respect to the sun is reflected in a westward direction then the westward movement of the earth in its orbit relative to the sun increases the photon's velocity relative to the sun but provides that photon a velocity of c relative to the earth. When a photon from the sun is reflected in an easterly direction relative to the earth that photon is still on balance reflected in a western direction relative to the sun, is given an increased velocity relative to the sun, and is given a motion of approximately c with respect to the earth. When a photon from the sun is reflected in a northerly or southerly direction by a surface on the earth the velocity of that photon continues at c with respect to both the sun and the earth.

Relative to the earth's surface the velocity of a reflected photon can also be affected by the velocity of the reflecting earthly body relative to the earth's surface. But the movement of bodies on the earth is rarely large relative to the high photon velocity, so the apparent velocity of reflected photons on the earth normally appears to be approximately c with respect to the earth. As a result when near the earth two photons are traveling in a particular direction side-by-side they usually appear to continue in that direction side-by-side even though the photons may have originated from sources which were moving to a small extent with respect to each other. When two such photons are directly approaching each other in opposite directions the distance between them seems to be reduced at a rate of approximately 2 x 300,000 kilometers per second, that is at 2c. When two photons approach each other by perpendicular paths the photons seem to come together at a rate of 1.414c. If a positron going just north of east, for example, joined an electron going the same speed just south of east then the resulting photon would continue eastwards with a velocity equal to the earlier eastward velocity of its components.[8]

Inside buildings many of the photons experienced originate when electrons comprising an electric current, to be discussed later, flow rapidly through a substance, such as the wire in an incandescent light bulb, causing photons which had been attached to components of the wire to be knocked out. Such ejected photons also usually have a velocity of approximately c, which they probably had before becoming attached.

When electrons have one or more photons attached to, or orbiting, them the combinations have greater mass but the same net charge as the electron by itself would have. Such combinations are often also referred to as electrons but in this document will be referred to as electron/photon combinations.

Protons

There is no accepted explanation why there are no stable combinations intermediate in size between photon-electron combinations and the next larger combination. The next larger stable combination is a much larger and long-lived combination known as a proton with a net positive charge of one and a mass of approximately $16,726 \times 10^{-31}$ kg. The proton thus has a net charge equal in size but opposite in type to the electron even though the proton is far larger than the electron. Each proton may be composed of associated positrons and electrons, containing one more of the latter than of the former. It has not been possible, however, to break up protons so as to study their internal configuration in detail, so that the proton may be just a combination of clumps of positive and clumps of negative substance. The observed reactions of particles bounced off protons do suggest, however, that the components of the proton are not closely packed or uniformly distributed throughout the combination. These components could be held together by the same Coulomb influences which hold other combinations of electrons and positrons together. A large proportion of the volume of a proton appears, however, to be just empty space. Since it appears that electrons rarely become closely attached to protons it may be that somehow there is a greater concentration of positrons in the portion of the proton combination closest to its center with a "skin" predominantly of associated negative electrons toward the outside. The repulsive force of this negative "skin" usually prevents any electron approaching a proton from getting closer than some extremely small distance from the proton even though the electron when further away would be attracted by the net positive charge of the proton. Yet a proton does sometimes combine closely with an extra electron and a neutrino in the formation of what is known as a neutron. And since the proton sometimes appears to emanate influence over a short distance in some direction different than just the effect of the proton's overall net charge it seems likely that the internal structure of the proton is not symmetric.[9]

Neutrons

The combination of a proton, an electron, and a neutrino has been given the name of neutron in view of the fact that the combination has no net charge. This combination is sometimes broken off from other combinations but a neutron usually has a lifetime measured in minutes when it is not a part of some larger combination. In this case the proton, the electron, and the neutrino soon fly apart.

Nuclei

Combinations of protons and neutrons constitute the central core of the basic material building blocks known as atoms. This core, known as the nucleus, always has a net positive charge whose magnitude is determined by the number of positive positrons in excess of the number of negative electrons in the combination. In the nucleus of hydrogen there is just one proton. In the nuclei of elements larger than hydrogen there are always both protons and neutrons. But regardless of the number of neutrons in its nucleus, an atom is identified as an element of a particular type by the size of the overall net positive charge in its nucleus, that is by the number of its protons not counting those in the neutrons. This size is called the element's atomic number. Most atoms occur in several forms each having the same net charge but different numbers of neutrons. These different forms are called isotopes of the atom. An isotope of an atom of one type may have approximately the same mass as an isotope of another type. To reduce confusion atoms and isotopes are often referred to by their symbols with preceding superscripts indicating the total number of protons and neutrons involved and with postscripts or subscripts indicating the net positive charge of the nucleus. $^{56}\text{Fe}_{26}$, for example, refers to an isotope of iron whose nucleus has 26 protons plus 30 neutrons and a net charge of 26.

Nuclei having net charges from one to over a hundred have been observed. Of these those having net charges up to 83, the atomic number of bismuth, are generally stable for very long periods of time. Larger combinations gradually break down into smaller nuclei. These larger nuclei are called radioactive. It is not possible to predict accurately the length of time before a particular large nucleus will break apart, but it is possible to estimate the time during which half of a sample will disintegrate. Such half-lives vary from a matter of a small fraction of a second to millions of years. Observed nuclei-type combinations heavier than uranium, with its atomic number of 92, have been created and observed only in laboratories and those combinations existed only for small fractions of a second.

Sometimes when a nucleus has a violent collision with another combination there is what is called a nuclear reaction, which alters the atomic number of some of the atoms involved. A particularly important reaction of this was discovered in the late 1930's when it was found that when one of the uranium isotopes $^{235}U_{92}$ is bombarded with a neutron the atom can sometimes be made to break apart into two smaller atoms. The more abundant isotope of uranium, $^{238}U_{92}$, was found not to split, or fission, when bombarded, but if a sufficient quantity of the lighter $^{235}U_{92}$ isotope could be separated and brought together in a mass of at least a critical size it was possible to produce an immense amount of fission. The first $^{235}U_{92}$ atom to break apart required bombardment by only one fast-moving neutron. In addition to two new smaller atoms that action produced three fast-moving neutrons, which caused fission in three more $^{235}U_{92}$ atoms if they were present, and so forth, rapidly creating an explosion of neutrons having great destructive effect in total. This chain reaction was the process involved in the Hiroshima atomic fission bomb.

When appropriate absorbing material is placed between $^{235}U_{92}$ isotopes the fission process can be controlled so that no explosion occurs and heat useful for creating steam for electric power generation can be produced.

It has also been found that when certain combinations of hydrogen isotopes are crashed together at very high speeds they can be transformed into heavier fast-moving helium atoms and positrons in an interaction known as fusion. When this transformation occurs in a gas of hydrogen isotopes most of the resulting helium atoms and positrons transfer momentum to the surrounding gas causing further interactions of the same type, while the other helium atoms and positrons transfer their momentum elsewhere in an extreme explosion. This process has been harnessed to produce explosions far more powerful than the original atomic bomb, but it has not yet been possible to use the process commercially for the production of useful power. In a sense fusion, which creates heavier atoms, is the opposite of fission, which creates lighter atoms. This fission is apparently a process which is taking place in the core of the sun.

Atoms

In an atom there are a nucleus and orbiting the nucleus, a number of orbiting electrons or electron/photon combinations equal, in number to the net charge of the nucleus. For many nuclei there also exist combinations having slightly more or slightly fewer orbiters than the net charge of the nucleus. These combinations are known as ions of the element involved.

While an atom has no net charge, the ions have small positive or negative net charges. The un-attached electrons of an atom may have one or more photons associated with them but, since these photons have no net charge, they do not affect the total charge of the atom.

In an atom the negative electrons and electron/photon combinations are usually in orbit, circular or elliptical, about the positive nucleus at various distances from the nucleus. And both the orbiting bodies and the nucleus are normally also spinning about axes within themselves. In an atom the orbiting negative electrons and electron/photon combinations are attracted by the net positive nucleus but the orbiters rarely crash into the nucleus. The orbiters may pass close to a nucleus but are usually are held away from the nucleus by the centrifugal effect of their linear velocity in directions roughly tangential to the nucleus and, when any negative orbiter comes very close to the nucleus, probably also by repulsion from a concentration of negative influence in the outer portions of each nucleus. Orbiting electrons and combinations in an atom are influenced, moreover, both by the attraction of the positive nucleus and by repulsion from any other negative orbiters of the atom.

In view of the small mass of electrons and photons in comparison to the mass of protons the total mass of an atom is determined primarily by the mass of its nucleus. That mass is the same wherever the atom is. The weight of an atom, or of any body, when defined as the magnitude of the gravitational force pulling it toward the center of the earth, varies, however, with distance from the center of the earth. Accordingly the weight of an atom, or other body, is less when it is on top of a mountain than when it is at sea-level. Sometimes the mass of an atom is expressed, not in terms of units such as kilograms, but in what are called atomic mass units, which are multiples of one twelfth of the mass of the carbon atom $^{12}C_6$.

The diameter of an atom is always a very large multiple of the size of its nucleus. As a result the nucleus and orbiting electron and electron/photon combinations take up only a small proportion of the space occupied by an atom. When a particle, or combination, from outside an atom flies into the space occupied by the atom there are various possible scenarios for what happens next. The incoming particle or combination may pass through without coming close enough to the nucleus or any of the orbiting electron and electron/photon combinations to cause any discernible change in the path, velocity, or spin of any of the components of either the incoming body or the atom. The incoming body may come close enough to some component of the atom to alter the path, speed, or spin of the incoming body but without sufficient effect either to cause the incoming body to stay in the atom or to cause any part of the atom to break away. On other occasions an incoming body may knock loose some piece of an atom. And the incoming body may simply become attached to some part of the atom or enter into an orbit about that part of the atom.

Two cases of incoming photons have been of particular interest to physicists. In one case a rapidly revolving incoming photon collides with an electron in an atom and causes that electron to fly out of the atom in what has been named the photo-electric effect. In this case charge flows out of the atom to some other place in the magnitude of the negative charge of the departing electron. As a result the atom becomes a positive ion.

In the other case an incoming photon enters into an orbit about an electron which itself is orbiting a nucleus. The resulting electron/photon combination then has more mass but only the same net charge that the electron alone had previously. The combination feels, therefore, immediately after the collision the same Coulomb attraction toward the nucleus as the electron alone had felt earlier, but that attraction is no longer sufficient to hold the heavier combination in its previous orbit, and the combination moves further away from the nucleus. The electron/photon combination might then have enough momentum to escape from the atom, but in other cases the continuing attraction of the nucleus would gradually slow down the combination until it settled down into a continuing orbit further out from the nucleus on average than the electron's previous orbit. It can be calculated that for a particular electron/photon combination to have a continuing orbit the product of its mass, distance from the nucleus, and linear velocity squared must equal a certain constant, so that when the mass and distance from the nucleus increase the velocity squared has to decrease. In some cases these further out orbiting electron/photon combinations have long lives, but in most cases they come apart or are knocked apart in a much shorter time period. The less massive electron is then pulled back by the nucleus into a closer-in orbit, while the photon flies off.[10]

When the mass of an electron is combined with the mass of an associated photon, as described above, if the new combination does not altogether escape from the influence of the nucleus, there are various new combinations of velocity and distance from the nucleus which could theoretically provide a new repetitive orbit. But only one new size orbit is normally created. As the combination increases its distance from the nucleus the velocity of the combination is reduced by the continuing Coulomb pull of the nucleus as a function of the distance from the nucleus. When the increasing distance from the nucleus has reduced the tendency of the electron/photon combination to fly away down to the point where that tendency is just offset by the reduced Coulomb force the particular orbit existing at that particular point is continued. The effect of the incoming photon is sometimes described therefore as being "quantized." If two or more photons had become attached to the electron at the same time the resulting new orbit would be even further out from the nucleus. When an electron is orbiting with no photons attached the atom is sometimes said to be in its ground state. When there are one or more photons attached the atom is said to be in an excited state. When an electron with one attached photon in an excited atom loses its electron the Coulomb force pulls the electron back into a ground state orbit closer to the nucleus.

For the electron or electron/photon combination in a particular state, ground or excited, about a particular type of atom only an incoming photon with a spin in a narrow range of frequency will move into orbit about that electron or electron/photon combination. When a number of photons break away at about the same time their common frequency of spin can be perceived by an eye or other device and provide to the viewer the impression that the atoms which are the source of those photons are of a particular color which is a characteristic of the substance involved. When the photons striking a substance do not include photons of the frequency capable of entering into orbit of the electrons of that substance then that substance will not project its characteristic color. Photons of various frequencies may, however, bounce off the substance and give a viewer the impression that the substance is of the color of the frequency of those incoming photons.

It has never been possible to measure the exact orbit of a particular electron in a specific atom. Theoretically one might try to bounce a photon off a chosen particular electron and then try to determine the course of the electron by measuring the direction and speed of the recoiling photon, but even if those attributes of the photon could be determined it would not be known which of various possible combinations of direction and velocity in the electron had caused the observed motion of the photon. And in any event bouncing the photon off the electron would have changed the velocity and direction of the chosen electron. It has been possible, however, for particular atoms to prepare estimates of the likelihood of finding an orbiting electron at any time at specific distances from the nucleus. The most likely distance has been found to be that distance at which the orbiter, with its specific mass and velocity, would have a circular orbit. There is not a precise diameter for atoms of a particular type, however, since the orbiting electrons in that type of atom could have different elliptical paths.

Particular patterns have been observed for the orbits of atoms with different numbers of electrons. In hydrogen with its single electron the orbiter moves about the nucleus roughly in a plane. The planes of the electron orbits in hydrogen are not all the same, however.

In helium the two electrons remain in approximately a single plane with respect to their nucleus while orbiting in the same direction—so as not to run into each other—and keeping at all times an equal distance from the nucleus on opposite sides of the nucleus as a result of the mutual repulsion of the electrons. The helium orbiters could, however, orbit in the same direction as their spin or in the opposite direction. Each orbiting electron would feel attraction from the two positive charges of the nucleus and, at the same time, some repulsion from the more distant other electron on the other side of the atom.

For larger atoms most electrons orbit in pairs, like the helium pair, with the atoms in each pair at about the same distance from the nucleus on opposite sides. These orbiting pairs are referred to as orbitals. There are never more than two electrons at the same distance from the nucleus. The two electrons in an orbital are normally spinning about axes internal to the electrons. The influences passing between these two electrons from the negative components in the electrons usually induce such electrons to have spins in opposite directions but both orbit their nucleus in the same direction. That direction can be in the same or opposite direction to the direction of the spin of the nucleus itself.

Single electron orbits at a particular distances also exist. In the smaller atoms these single electron orbit distances exist only for the outermost electron of an atom having an odd number of electrons, but for some larger atoms there are sometimes more than one single electron in intermediate orbit as a result of complex interplay of forces among the nucleus and the many orbiting electrons. And the electrons in orbitals of atoms combined in larger bodies are often affected by the electrons of the other components of the larger bodies so that the electrons in the orbitals may not be symmetrically distributed about their nuclei.

In lithium when its three electrons are in their ground state two of them form an orbital while the third is pushed further from the nucleus in a single orbit which keeps it equidistant from the repellent equal negative charges of the inner two electrons. There are, however, two different types of orbit which meet this condition. The outer electron can either rotate about the nucleus in the same plane as the inner electrons relative to the nucleus or the outer electron can still keep its constant distances from the other electrons and the nucleus in an orbit perpendicular to the plane of the orbits of the inner electrons.

Despite the fact that the nucleus of the lithium atom has a larger total positive attractive force than the nuclei of hydrogen and helium, the repelling of the outer lithium electron by the other two electrons causes lithium to have a larger diameter than the two atoms with smaller nuclei.

In larger atoms the forces among the orbiters result in even more complex relationships among the planes of the various orbits.

Among the twenty smallest atoms there are six, helium, beryllium, neon, magnesium, argon, and calcium, which have all their orbiting electrons in two-electron orbitals.

Generally most likely to interact with other atoms are those atoms with outer electrons orbiting singly. Among these are carbon, nitrogen, oxygen, and sulfur.

The paired electrons in each orbital both usually take the same amount of time to complete one orbit around the nucleus, but the time period for one revolution varies from orbital to orbital. Since the attraction of the nucleus is less for those electrons further away the average linear velocity for far out ground state electrons is less than the velocity for ground state electrons closer to the nucleus. The time for one complete revolution around the nucleus is therefore longer for ground state electrons further out both because of their slower speed and because of they have further to go for one revolution. Although an orbiting electron/photon combination further out may have less velocity than an electron closer to the nucleus the outer combination could have greater momentum since its mass would be greater.

When an incoming photon approaches an orbiting electron when the velocities of the two bodies are not too dissimilar the photon may be attracted into orbit about the electron if the rotation of the photon is at a rate which keeps a positive face of the photon toward the negative electron at all times as the electron orbits its nucleus and the photon swings around the electron. For this to happen the appropriate rotation of the photon will depend on the difference between the linear speeds of the electron and the photon, When a photon, after having been in orbit about an electron, breaks away or is knocked away observation of the rotation of that photon can tell something about the atomic element which had been involved. It has been possible to catalog the rotation of out-going photons from each type of atom, so that it is now possible to determine from the frequency of the rotation of photons emitted what type of atom was involved.

The different types of atoms are often displayed in a rectangular structure of rows and columns known as the Periodic Table of the Elements. In one form of the table the further an atom is listed first to the left and then up the table the larger the number of protons in its nucleus. In most cases the atoms in a particular column have the same situation for their outermost electron, that is either a single orbit or participation in a paired orbit. In general, the atoms in a particular column exhibit similar tendencies for various physical properties, such as melting points, and for participating in chemical reactions, as discussed below.

Molecules

Without altering their nuclei atoms can join with other atoms to form larger combinations known as molecules. Changes such as these are known as chemical reactions. Molecules can be composed of atoms of the same type or of different types. Formation of molecules can only take place when the atoms are not moving too rapidly with respect to each other when they come together, and a particular type of atom may combine with certain other types and not with others. The number of atoms in a molecule may be as little as two or as many as thousands. When atoms come together they may simultaneously form molecules of more then one type. Sometimes particular atoms will join together only if heated, and sometimes molecules break apart into their constituent atoms when heated.

The basic process by which two atoms are held together in a molecule is that positive nuclei in different atoms exert their attraction on the same orbiting electron or electron/photon combinations thus indirectly creating a bond between the nuclei. In some cases the particular orbiters involved in the bonding continue to orbit the same nucleus they orbited before the molecule was formed. In other cases an orbiter may continue to have a relationship with the same nucleus as before but will have transferred its orbit to a different nucleus when the molecule was formed. And in the case of some metallic molecules the relevant electrons and electron/photon combinations form a sort of sea of particles with no unique attachments to particular nuclei.

Generally those atoms whose outermost orbiter is not part of a pair are more prone to form molecules than those atoms whose outermost orbiters are a pair, and those atoms whose outermost orbiter is shielded from the positive nucleus by more inner orbiters are more likely to have an orbiter drawn away into orbit around the nucleus of another atom.

Sometimes when a body largely composed of atoms whose outer electron or electrons are not strongly attracted toward the nuclei touches another body whose atoms have a strong attraction for electrons then when the bodies are pulled apart some electrons from the former body stay with the latter body. In this case the former body becomes positively charged, the latter body becomes negatively charged, and the bodies then have an attraction for each other. This happens, for example, when silk cloth and amber are brought into contact, and the effect is intensified if the two materials are rubbed against each other to increase the amount of physical contact.

Evanescent Combinations

In addition to the combinations discussed above, which are the basic building blocks of matter as we experience it, a large number of evanescent combinations have been observed with lives in each case of less than a very small fraction of a second. Generally these short-lived combinations have been created in the laboratory though a few have been observed to be created by collisions with the earth's atmosphere by combinations flying in from outer space.

States of Matter

Aggregations of the various non-evanescent combinations discussed above can exist in five different ways or states: solids, liquids, gases, plasmas, or condensates.

A solid is formed when the same processes by which atoms are held together to form molecules serve to hold atoms in different molecules together. As in single molecules the strength of the force binding the parts of a solid together varies over a wide range depending on the types of atoms involved. A solid may be composed of molecules of the same type or of molecules of various types. When the molecules are held together in some definite repetitive pattern the solid is known as a crystal. In a solid each nucleus maintains roughly the same position relative to other nuclei nearby, and the aggregation can maintain a three-dimensional shape even though not constrained by a container, but the nuclei nevertheless normally vibrate, and sometimes rotate, with respect each other to some extent.

A liquid is formed when the molecules in an aggregation are generally attracted to each other with a force which is not sufficient to keep the molecules from moving with respect to each other but which is sufficient when combined with the force of gravity to keep most molecules from flying up and out of an open-topped container. In most liquids, however, some molecules do break away in what is known as evaporation.

In a gas the molecules are moving with such vigor with respect to each other that the aggregation can usually be held together only by a container which is closed in all directions. In the container the molecules are flying about frequently colliding with each other and with the walls of the container. Our atmosphere is a special case in which the gas, our air, is not constrained by a container with a lid. In this case the molecules are so diffuse that they are ultimately prevented from flying off into space by the continual gravitational attraction of the earth.

Most types of atoms are in the solid state at some low temperature. Above a certain temperature, known as the freezing point, they become liquids. Above a certain higher temperature, known as the boiling point, they become gases.

At an extremely high temperature electrons and nuclei fly about rapidly with no fixed relationship with each other. Combinations in this state are said to be in a plasma.

There have been preliminary reports that it has also been possible in a laboratory to cool some substances to such a low temperature that their protons and neutrons broke apart into smaller combinations and possibly into their constituent electrons and positrons in what is known as a Bose-Einstein Condensate.

Gravity

When the motor dies on a plane, the plane eventually falls to the earth. The influence which brings about this result, known as gravity, is acting in this case between the earth and the plane but the same influence is operative between any two atoms or larger bodies. Gravity is conventionally considered to be a very weak influence entirely separate from Coulomb influence. Yet the influence on the plane, like Coulomb influence in the examples discussed above, varies inversely with the square of the distances between the bodies involved and gravitational influence also travels at the velocity of c relative to its source. Consistent with these facts the model assumes that gravitational influence is not a separate type of influence but just a manifestation of Coulomb influence in a certain type of situation even though the gravitational force of attraction between two molecules with no net charge and no magnetic orientation is only on the order of 1×10^{-34} of what that strength of attraction would be for bodies of the same mass if one body were composed entirely of electrons and the other were composed entirely of positrons.

The gravitational attraction between uncharged bodies may arise because the force of Coulomb repulsion between two particles of the same type may be just slightly smaller than the force of attraction would be if the articles were of the same type. Unfortunately the strength of gravity is so very small that it has never been possible to measure and compare the forces of attraction and repulsion with sufficient precision to test this hypothesis. An alternative which the author judges more likely is that gravitational attraction results from interference with the movement of Coulomb influences by atomic components in the path of attracting and repelling influences moving between pairs of particles in the bodies involved. A somewhat similar hypothesis was put forward by a friend of Newton's, Nicholas Fatio de Duillier and was advocated in the 1930's by an Italian physicist Quirino Majorana. .

When two combinations without net charges are apart, each particle of each combination sends a portion of its Coulomb influence toward each particle of the other body. Depending on the arrangement of the particles, a particular combination could be either attracted toward the other combination or repelled. In practice when the combinations have no net charges, and are not very close together, the positive and negative forces seem to offset each other and there appears to be neither appreciable net attraction nor repulsion. But that is probably not the case when at least one of the bodies is massive, that is composed of a very large number of atoms. For a simple illustration consider two illustrative atoms some distance apart each with a nucleus with a positive charge of two and two orbiting electrons each with a negative charge of one. If at some moment no component of either atom was in a position to block the flow of influence between any other components then each component of, say, the right atom would be exerting either an attraction and repulsion effect on each component of the left atom. An electron of the right atom would contribute to a repulsive force of, one divided by the square of the distance involved, on each of the two electrons of the left atom while contributing to an attractive force, of two divided by the square of the distance involved, on the nucleus of the left atom. These forces would not be exactly offsetting to a minuscule extent as a result of the extremely small differences in the distances the influences would have traveled when one or both of the atoms had their components lined up along a line drawn between the atoms. On such an occasion, in view of the fact that the shortest distance involved would be between the two closest electrons repelling each other, it can be calculated that the atoms should actually slightly repel each other on balance. But to some very small extent the components of atoms probably do sometimes get in the way of the passage of the influences, and the effect of this interruption would more than offset the periodic minuscule repulsion otherwise existing between atoms. Probably the most significant interference with the flows of influence to an atom on the left from an atom on the right would occur during the time when one of the electrons on the right was to the right of its nucleus, which is far more massive than an electron. As the electrons in the right atom revolved about the nucleus some of the time the right nucleus would create to the left a shadowed area from which influence from the currently right-most electron had been blocked by the nucleus.

The portion of influence which had been blocked would have on balance repelled the left atom. In the absence of this influence the left atom would on balance be attracted to the right. The attraction would be very small from each pair of atoms involved but trillions of atoms could be involved. In this way, for example, there could be gravitational pull on the moon from the earth and on an apple in a tree from the earth.

There could also be a slight blocking effect when, for example, an outer electron of a combination on the moon blocked an incoming attractive influence from the earth from reaching the positive nucleus of that combination. This effect would be very small.

Similar interference with the transmission of weak Coulomb influences probably also explains why it has been observed that the precession of a pendulum on the earth speeds up when the moon passes between the earth and the sun.

Temperature

The more the components of a body are moving about with respect to each other the greater is said to be the temperature of the body. The temperature would be at its lowest possible level when there was no motion at all of the components of the body, though it has never been possible to cool a body to quite that level. Temperature can be measured in various ways, including by observing the extent to which heating increases the ease with which an electrical current flows through the body, expands the size of a body, or increases the pressure exerted by a gas or liquid on its container. The second of those methods is employed in the household thermometer in which temperature is measured by the extent to which heating increases the volume of the liquid in the thermometer. Precise measurements are sometimes attempted by noting the extent to which heating a gas increases the pressure of the gas on some portion of the container for the gas. Heating the gas increases the frequency and velocity with which molecules in the gas collide with the container.

There are in use three principal scales for indicating temperature. On the Kelvin scale zero represents the absolute theoretical minimum temperature for a substance. The freezing point of water is at approximately 273.15 degrees (°) Kelvin (K) and the boiling point of water is at approximately 373.15°K. For the Celsius scale the degree covers the same temperature range as in the Kelvin scale, but in the Celsius scale all temperatures are set 273.15 degrees lower. In the Celsius(C) scale the absolute minimum temperature is considered to be -273.15°C, the freezing point of water is at 0°C, and the boiling point of water is at 100°C. The Fahrenheit scale is set at 9/5 of the Celsius scale plus 32 degrees. In the Fahrenheit scale the minimum temperature is 460.3°F, the freezing point of water is 32°F, and the boiling point of water is 212°F.

The temperature of a body can be raised in various ways. When a body is bombarded by photons those incoming photons can accelerate the components of the body, thus raising its temperature. If all those photons bounce away the body's temperature will be raised while its mass is left unchanged. But if some of the incoming photons remain with the body then both the mass and the temperature of the body will be raised. Increasing the temperature of a body will not raise its mass unless particles, atoms, or molecules are added to the body. By another procedure when a substance whose atoms and molecules have on average greater momentum—and therefore higher temperature—is brought into contact with a substance with lesser momentum and temperature then the temperature of the portion of the warmer substance near the point of contact is lowered and the temperature of the portion of the cooler substance near the point of contact is raised. Gradually in this case the cooling influence spreads outward in the warmer substance from the point of contact, and warming influence spreads outward in the cooler substance from the point of contact. The pace of the spread of temperature change is not the same in all substances, however. Temperature change spreads very rapidly, for example, in copper but very slowly in water. The average temperature of a substance can be lowered, not only by contact with a cooler substance, but also when some of the higher momentum, that is warmer components of the substance, fly away by evaporation and when the pressure on a gas is reduced.

Waves and Sound

The transmission of influence from one body to another can involve the movement of intermediate bodies. For example, suppose a number of similar balls all with net negative charges were lined up in a row at rest with respect to one another not touching each other and the first ball in line were hit toward the others. As the first ball neared the second ball the first ball would eventually come to a halt and increased repelling force would push the second ball toward the third, and so on until the last ball in line were given a shove. In this case it would be said that a single longitudinal wave had moved along the row of balls in the direction the balls had moved. If the last thing in the line were not a ball but some massive object like a wall the transmitted influence would give that last object a very small movement in the direction of the influence force while at the same time causing the last ball to recoil back in the direction from which it came. The recoil influence would be a bit less than the original but would be transmitted all the way back until the original ball hit were pushed in the direction opposite to the original influence. If then the original ball were than hit again in the original direction, and so forth, the balls in the middle would end up moving back and forth along the line of transmission of the influence with no net movement of those balls.

If instead there had been a lot of balls of different sizes touching each other in a horizontal trough what are called transverse waves could have been created. A push on a large ball at, say, the left end of the trough would not primarily push it immediately directly to the right. Rather that ball and some other balls near the left end of the trough could, however, be pushed somewhat up and to the right and be piled up in a mound or ridge. Then the pull of gravity on the balls up in the ridges would pull them downward back to their original positions all the while pushing other balls to their right up into other ridges. A ridge would appear to move along the trough. If the balls at the left end were given a series of pushes a series of transverse waves would be created.

Waves which appear on the surface of a body of water are primarily transverse waves. When a pebble is dropped into a pond the resulting ridging up of parts of the water can be seen in the ripples which fan out from the point where the pebble was dropped in. When such a transverse water wave comes to a barrier which has a partial opening to more water beyond the barrier the parts of the waves hitting the barrier reflect backward while the other parts of the wave pass through the opening. The parts of the waves passing through the opening do not, however, just continue in the original direction. The parts of the wave in the opening act as if they were an entirely new wave source, and new waves move off in all directions on the water surface beyond the opening.

In practice waves usually encountered are to some extent combinations of the two different types just described. Sound is, however, transmitted primarily by longitudinal waves. If the end of a metal bar is struck by a hammer the molecules which are struck are pressed toward the next set of molecules in the bar, and that set is pressed by electric repulsion and possibly also by physical contact toward the next set, and so on in all directions from where the blow fell on the bar. Each set of molecules recoils when it hits the next set, so the molecules vibrate for a while. Compression and then recoil of the molecules thus passes to the other end of the bar where the vibration could be felt after a certain time interval which depends on the type of metal in the bar. If the hammer strikes repeated blows in rapid succession continued vibration of a certain frequency would be detectable at the other end of the bar by touch and possibly also by hearing if the frequency of the bar's vibration were high enough and not smothered by the effects of other vibrations. Similar compression, recoil, and vibration are caused when a flat surface is moved rapidly and repeatedly against air, for example by a musical tuning fork. In both the case of the hammer against a bar and the case of the tuning fork against air the process is the same, and it is said that sound has been transmitted in both cases even though most of the sounds we deal with are transmitted through air. In both cases the transmission involves movement of physical bodies and not just action-at-a-distance influences. A tuning fork vibrating in the center of a perfect vacuum would transmit no sound since there would be no molecules to convey the sound.

The intensity of sound perceived depends upon the momentum and area of the surface which originally compressed the transmitting molecules, but the speed of the sound from source to listener varies primarily not with the intensity of the compressing influence but rather with the nature of the transmitting material. The speed with which the molecules in a material compress and recoil is a characteristic of that material whatever the violence of the action at the source. And the speed of sound does vary markedly among different materials. In iron the speed is about 5,130 meters per second. In air at 0^0 C the speed is about 331 meters per second. A listener at a sound traveling through air would not notice any difference in that speed depending upon whether the source of the sound were moving slowly or rapidly toward the listener. But if the source were moving toward him—whether directly or at an angle—there would be a shorter time interval between the arrival of one compression and arrival of the next compression. He would hear a higher frequency of sound. If the source of the sound were moving away he would hear a lower frequency of sound. This variation in frequency is known as the Doppler Effect. It is the explanation for the fact that the sound from an approaching ambulance sounds shriller than the sound from an ambulance moving away.

For the human ear the audible range of detectible frequencies is from about 20 vibrations per second to about 20,000 vibrations per second. Air compressions arriving at a human ear at the same time from separate sources can augment the intensity of the sound heard. When compression from one source arrives at just the time when rebound has minimized pressure from sound from another source the two sounds can interfere with each other reducing the intensity of the sound perceived and perhaps even eliminating it altogether.

Magnets

Bodies composed of certain materials found on the earth, notably a form of iron known as magnetite, sometimes attract or repel each other depending on their orientation with respect to each other. These bodies are known as natural magnets. A magnet of this type can induce magnetic attributes of the same type as those of the natural magnet in some other types of material when the magnet is placed near the other material. Some materials, such as iron, are particularly susceptible to being made magnetic if they are heated to a certain extent. Depending on the material involved it is possible in this away to produce permanent magnets or to produce temporary magnets which lose their magnetic attributes when the strong magnet is taken away. The former process probably explains the existence of most natural magnets on or near to the surface of the earth. The material in these magnets was probably once heated and was then influenced into magnetic arrangement by the large strong magnet which exists deep within the earth. That earth magnet projects influences as if it has one end, known as a pole, now beneath a point north of Hudson's Bay in Canada about 1300 miles from the north geometric pole, and the other pole beneath a point a similar distance from the south geometric pole. These earth magnetic poles moves, but the movement is so slow that it is still possible to use the magnetic poles for navigational purposes after correction for the known current distances between the magnetic and geocentric poles. One end of a navigator's compass, which is just a freely floating magnet, points toward the earth's magnetic north pole. This end of the navigator's compass, and of any other magnet other than the large earth magnet, is known as the north-seeking, or more commonly the north, end of the magnet. One of the ends of a magnetized needle, when allowed to, also points down toward the nearest magnetic pole of the earth, and the extent of this "dipping" can be used to estimate the latitude of the needle at the time.

Despite the fact that heat up to a certain level can facilitate the creation of a magnet all bodies lose their magnetism when heated beyond a certain point.

When two rectangular bar magnets are positioned so that the north pole of one is near the south pole of the other, and the other poles of those magnets are far apart, the two magnets attract one another, as shown on the upper portion of **Diagram 2.**

Diagram 2

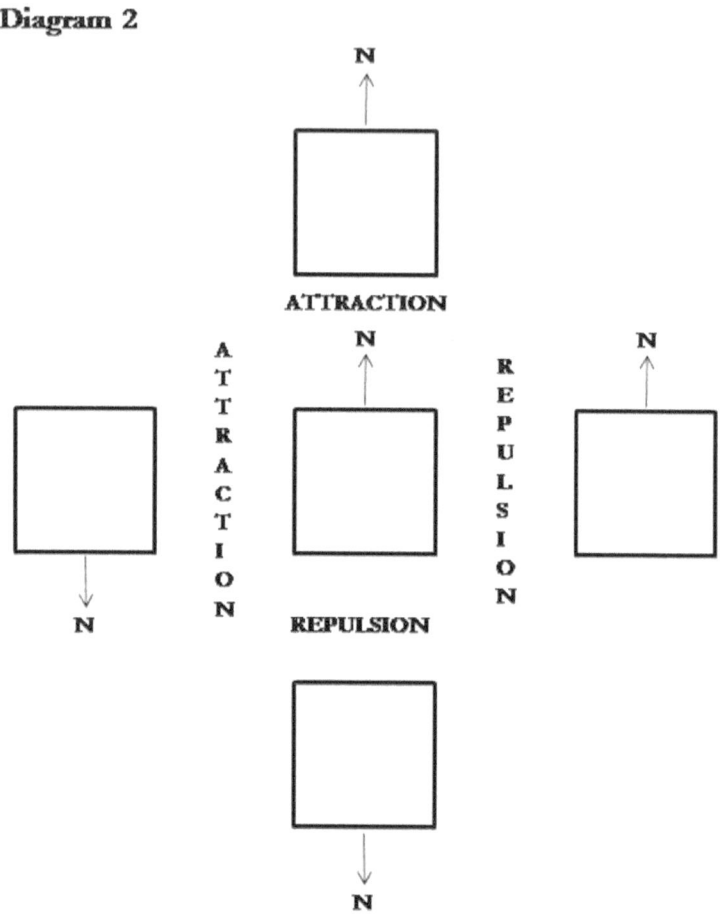

An N adjacent to each of the magnets in the diagram indicates its north-seeking end. This attraction remains the same if either or both magnets are rotated about a line running lengthwise from pole to pole through the magnets. If the magnets so positioned are large and a considerable distance apart the strength of the attraction is approximately inversely proportional to the square of the distance between the midpoints between the poles of the magnets. If the magnets so configured are small and close together the strength of attraction varies approximately inversely by the cube of their separation. If the north—or south—poles of two magnets are placed close to one another with the other poles far apart the magnets repel each other with a strength which varies over considerable distances approximately inversely with the square of the distance between them. If the two magnets are placed side by side with the north poles close to each other and the south poles close to each other the magnets repel each other with a force which varies inversely with the square of the distance between the magnets, and if in this configuration the north pole of one magnet is placed close to the south pole of the other magnet while the other south and north poles are close together then the magnets attract each other. For other positions of the magnets with respect to each other the forces on the magnets are intermediate between those just described.

A more general display of the relationships between magnets is shown below in **Diagram 3**, which shows a central bar magnet surrounded by some other smaller bar magnets each of approximately the same strength of net attraction from the central magnet.

Diagram 3

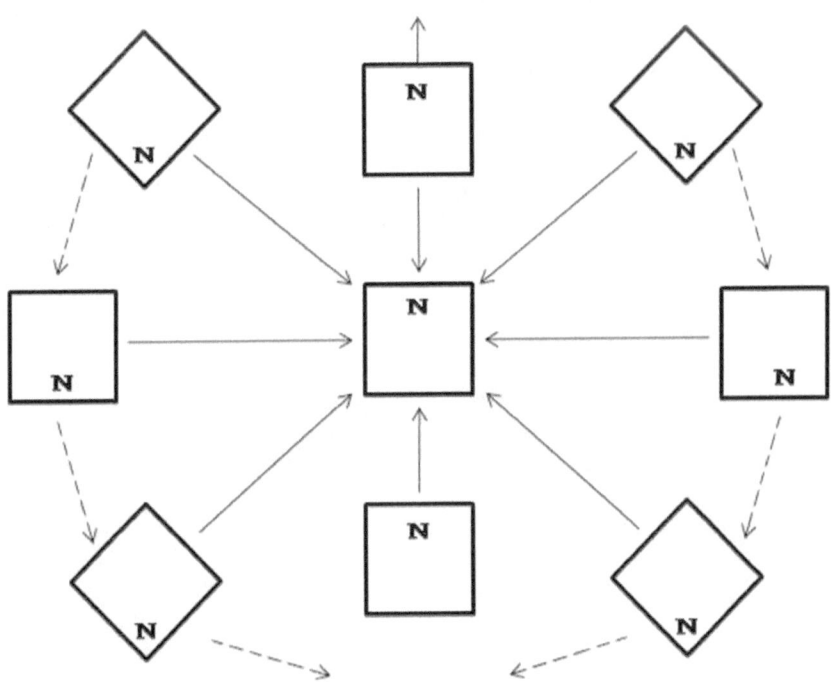

The direction of the attraction for each small magnet is shown by the line of force arrow from the small magnet to the larger magnet. Dashed arrows indicate the direction of the north ends of the small magnets at various positions relative to the central magnet. Each of the small magnets is shown with the orientation with which it would not be experiencing any net pressure for change in orientation. If any one of the small magnets were twisted so that its north-seeking end were pointing in the direction opposite to that shown in the diagram that magnet would then experience repulsion.

The influences projected by a magnet are not a separate type of influence different from Coulomb influences but rather the projection of Coulomb influences in a particular pattern. Although most magnets consist of many molecules the origin of magnetic influences can be illustrated simply by reference to magnets composed of just two molecules which are so constructed that each molecule has an outer electron while the inner portion of the molecule consists of an aggregation of electrons and nucleus having in total a net positive charge of one. Two such magnets, one atop the other, are illustrated in **Diagram 4.**

Diagram 4

Viewed from above the North side

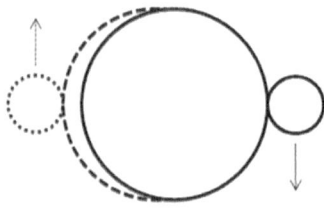

Viewed from the Side

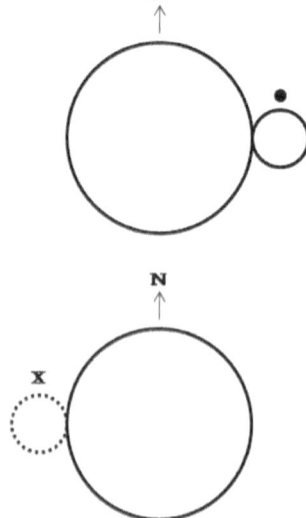

The upper portion of the diagram depicts the magnets as seen from a point above their north ends. The upper magnet is shown in solid lines, with the big circle representing its central portion and the smaller circle representing its outer electron. The lower magnet, shown with dashed lines, is offset a bit to the left to aid visibility. Both magnets have their north ends facing toward the viewer up out from the page. The arrows next to the electrons are intended to indicate that both electrons are orbiting around their central portions in a clockwise direction as seen from above.

The lower portion of the diagram depicts the magnets as seen from the side. The electron of the upper magnet has above it a dot, a symbolic representation of the point of an arrow approaching the viewer, to indicate that that electron is coming out of the page. That electron will next orbit around the central portion by moving horizontally to the left. The x above the lower electron, a symbolic representation of the view toward the feathered tail end of an arrow, indicates that the lower electron is moving into the page. That electron would then orbit on around the back side of the central portion.

In the case illustrated the electrons are assumed to be on opposite sides of the magnets and to be orbiting with the same velocity. The negative lower electron would attract the positive upper central portion and repel the negative upper electron. The positive lower central portion would repel the positive upper central portion and attract the negative upper electron. For the relative distances involved in this case there would be a net attraction between the two magnets in view of the effect of the long distance between the electrons repelling each other. If the electrons were on the same side of the central portions the magnets would repel each other, but apparently in real magnets the repulsion between the electrons keeps them on opposite sides so that magnets so configured do attract each other. If, on the other hand the lower magnet were made to have its north end point in the opposite direction from the upper magnet then the lower electron would orbit in the opposite direction and the two magnets would necessarily have their electrons on the same side half the time. In that situation the two magnets one above the other would repel each other.

If the two magnets were side by side both having their north sides pointing in the same direction up out of the page, as shown in the upper portion of the view from above in **Diagram 5**, the magnets would repel each other when their electrons were on opposite sides.

Diagram 5

Viewed from above the North side

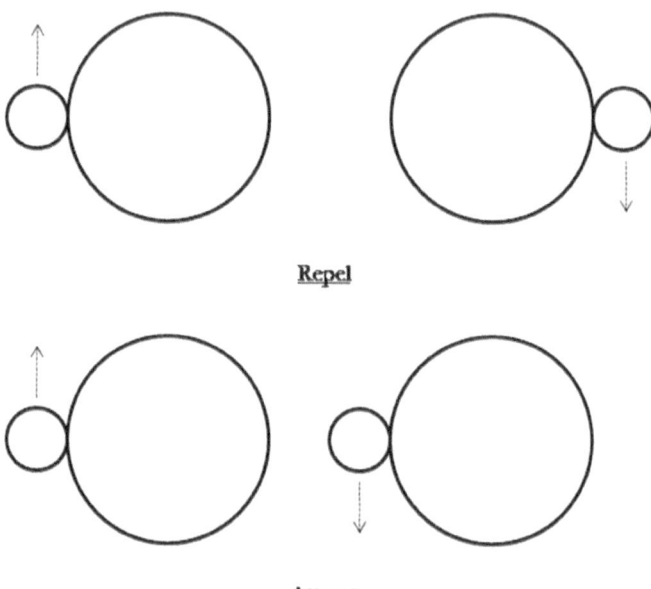

Repel

Attract

The repulsion between the central portions would not be fully offset by the attractive influences which have a longer way to travel between electrons and central portions. And whenever the electrons were in any other position relative to each other the influences passing between the electrons would tend to push them toward the positions displayed in the diagram. Side-by-side magnets would then normally repel each other when their north sides pointed in the same direction. If, on the other hand, the north side of one magnet pointed out of the page while the south side pointed in the same direction and the electrons were on the same sides of central portions, as illustrated in the lower portion of Diagram 5, the magnets would attract each other, and when the electrons were in any other relative positions the influences passing between the electrons would tend to push them toward the positions shown in the diagram.

In most magnets many more molecules would be involved, and it might be wondered whether the Coulomb influences passing between molecules, even when the bulk of them had their north sides in the same direction, wouldn't offset each other to a considerable extent and reduce the magnetic effect. It has been found, however, that, when there are a number of molecules side-by-side with their axes parallel and their north sides in the same direction, the effects of this configuration can be analyzed as if the electrons had all moved to the outside of the combination and were orbiting around the outside. As a result such a combination acts as if it were a single powerful magnet.

The molecules shown on Diagrams **2** and 3 could be just portions of larger aggregations in which the preponderance of the magnets had their internal axes about which the outer electrons were orbiting, pointing in the same direction. If such larger magnets were broken apart each part would still be a magnet.

Magnets of different materials would all project positive and negative influences in the same pattern as illustrated by the magnets discussed above, but magnets of different materials could have different molecular arrangements.

The effects shown on Diagrams 4 and 5 explain the effects of magnets on each other as a result of the orbiting patterns of the outer electrons of the molecules in the magnets and make clear why magnets interact with each other as they do. There are also, however, other important effects resulting from the coordinated nature of the spins as well as of the orbits

of the electrons of the molecules in magnets. These will be illustrated on
Diagram 6.

Diagram 6

Viewed from above the North side of the Magnet

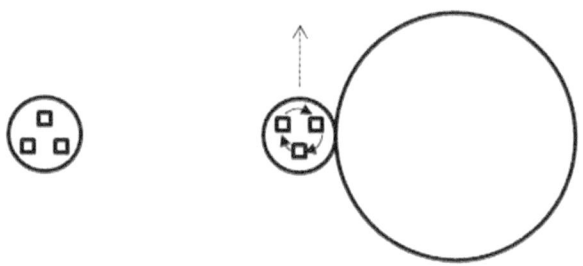

That page shows enlarged depictions of the central portion of a molecule
plus two electrons. The large circle is the central portion. The electrons are
the circles about the size of a quarter. The one on the left is free-standing.
The one on the right is an electron on the edge of the central portion of a
molecule of a magnet whose north side is vertically up off the page toward
the viewer. The view is from a point above the magnet's north side. In
the electron on the right the small squares inside the electron represent
just two out of many component sections of the electron. Each of these
sections has a negative charge. In a magnet these sections rotate about
an internal axis of the electron in the manner indicated in the diagrams.
As viewed from above the rotation is clockwise just as is the orbit of the
electron. As a result the electron on the left receives influences pushing it
toward the top of the page both from the spin of the elements of the right
electron and from the orbit of that electron about the molecule's central
portion. Clockwise spin of the molecule's central portion could also provide
some influence in the same direction but influence from the center of
molecules does not usually play a major role in the strength of a magnet. If
the left electron in the diagram were at rest with respect to the magnet the
influences reaching the left electron would push it directly toward the top
of the page. If the left electron had been moving from the left across the
page toward the magnet those influences would bend its path toward the
top of the page. In this way the magnet can cause movement in another
charged body even though the magnet has no net charge to pull or push a
charged body directly toward or away from the magnet. The effect of such
influences emanating from the spinning and orbiting of the electrons in

magnets are of great practical importance, as described in the following section concerned with electricity, even though the spins by themselves are not of significance in explaining the interactions between magnets themselves [11].

Electric Current

When a sequence of like-charged bodies, notably electrons, moves through a particular space an electric current is said to move through that space. When the movement through the space is in the same direction for a period the current is called direct. When the current rapidly changes direction back and forth the current is called alternating. The carriers of charge in a current can be attached to and carried by a moving body, such as a rotating disc. The individual charge carriers can also move directly through open space and through spaces in some materials, such as a copper wire.

The movement of individual charge carriers not attached to some moving body is induced in three principal ways: by repulsion from a concentration of like charges behind and attraction of a concentration of opposite charges ahead of the carriers, by a pair of certain types of chemical reactions, and by the effect of repeated radiation of a certain pattern of Coulomb force on to a wire. This repeated radiation is sometimes created by the relative physical movement of a magnet and a nearby wire, sometimes by the effect of an electric current in one wire on the components of another wire nearby, and sometimes by a series of revolving photons passing near or colliding with a wire.

The first of these methods for inducing a current, response to charge concentrations, is illustrated by a stroke of lightning. The lighted streak reveals the path of a stream of electrons passing down to the earth from a concentration of negatively-charged ions which have built up at the bottom of a cloud of water droplets and caused a net positive charge to build up on the nearest surface of the earth. Small flows of electrons called sparks can also pass between concentrations of opposite charge which have been built up in some pairs of materials, such as silk and glass, by rubbing them together.

When a material capable of conducting electrons is connected to two charged bodies a current will flow between the bodies, not only when the bodies have opposite charges, but also when the two bodies have unequal concentrations of the same type of charge.

In the second method of creating an electric current the chemical reactions involved often take place in a battery. In a common form of battery there are two rods or plates, called terminals or electrodes, of two different elements, for example zinc and copper, immersed some distance apart in a prepared chemical mixture. This mixture is chosen because it reacts with both terminals but in different ways. The mixture attracts electrons out of the kind of material of one terminal, leaving that terminal with a net positive charge, and deposits electrons on the material of the other terminal, leaving it with a net negative charge. When one end of a wire of, for example, copper, is attached to the negative terminal of the battery and the other end is attached to the positive terminal, some of the concentrated electrons at the negative terminal are pushed into the wire at the same time pushing some of the electrons in the wire along the wire. Simultaneously the positive terminal of the battery is attracting electrons out of the wire, making room for the electrons coming along the wire from the other end. In this way a flow of electrons, that is a direct current, is established in the wire. The chemical reactions in the battery continue for some time to transfer electrons from the positive terminal pole to the negative terminal, thus allowing the current flow in the wire to continue.

Not all materials from which a wire could be constructed provide the same conductivity for the flow of electric current. Metals are good conductors of electricity. Non-metals essentially do not conduct electricity at all, and so-called semiconductors have a conductivity somewhere between that of metals and nonmetals. Those materials, like copper, which offer the least resistance are generally those made from atoms which have one or more of their outermost orbiting electrons least firmly attracted toward the remainder of the atom. A discovery which turned out to have crucial significance for modern electronic devices was that certain materials which in their pure form have very little conductivity, notably silicon and germanium, could be converted into useful semi-conductors by the addition of a small amount of impurities. Switches, amplifiers, and devices for aiming projected electrons when made from these so-called solid-state semiconductors are more efficient and far less costly than the vacuum tubes previously used for such devices. In the vacuum tubes electrons projected through a vacuum were deflected by varying the charges in plates alongside the electrons' path through the vacuum. In the modern devices when certain types of semi-conductors are placed in contact with each other running a small electron current sideways through one of them, or exposing one to sunlight, results in a significant but controllable flow of electrons from one of the semi-conductors to the other and, when desired, around a wire circuit from one semi-conductor back to the other semi-conductor.

In a wire the moving charges are primarily some electrons which individually do not remain in orbit about the same nuclei. In a wire with a current flowing the positive ions from which some electrons have become separated move to some extent in the direction opposite to the average movement of the separated electrons but the mass of the ions is so great relative to the mass of the separated electrons that the distance moved by the ions is small relative to the movement of the unattached electrons. By very old convention the direction of an electric current is said to be in the direction opposite to which the electrons move. The moving electrons in a direct current in a wire do not, however, move in a straight line along the wire. They are deflected individually in various directions over time but their average net movement is in one direction along the wire. The electrons in a current-carrying wire have an average velocity along the wire of only millimeters per hour even though most of the individual electrons are usually moving in some direction at each instance with a velocity which is a substantial proportion of c. The influence of a current in one part of a wire in causing current in another part of the wire moves with a velocity of approximately 75% of the speed of light. As these current-carrying electrons move about in a wire they may also knock loose some attached photons, which then fly away from the wire.

When a wire which has a current flowing in it in a particular direction is brought near and parallel to a second wire in which previously no current was flowing, a current is induced to flow in the second wire in the particular direction. Apparently the moving electrons in the first wire push electrons in the second wire to move in the direction the electrons in the first wire were moving. And, of great practical importance, the effect on the second wire is increased if current from a single source is brought repeatedly past the second wire when the first wire is coiled so that many segments of the first wire are near the second wire.

When segments of two wires are side-by-side and currents have been made to flow in the same direction in both of them the segments attract each other. This attraction results from the fact that any electron in one of the wires passes close to a nucleus in the other wire—and is thus attracted—more often than that electron passes close to an electron moving in the same direction in the other wire—and is thus repelled. When the currents in the two wires are moving in opposite directions electrons in a wire pass close to electrons in the other wire more often than close to nuclei in the other wire and so the wires are repelled. This latter effect may also be intensified because electrons moving in opposite directions in the two wires probably slow each other down and increase the repelling concentrations of electrons in the two wire segments. [12]

An extremely important method of inducing an electric current involves moving a magnet and a wire closer to one another. The induced current results, however, not from that movement but from the fact that after the movement charged particles in the magnet and the wire are closer together and thus stronger Coulomb forces pass between the particles involved to push the electrons into a flow of current. An example can illustrate how this happens by reference to the **Diagram 7.**

Diagram 7

Viewed from above the Magnet's North Pole

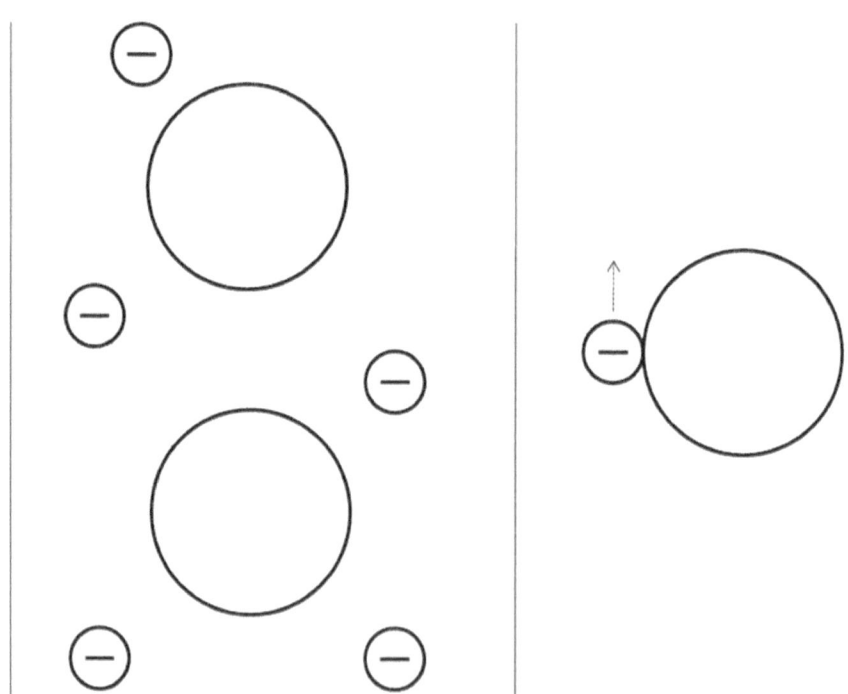

The right area of that diagram shows one molecule of a magnet whose north end points vertically out from the page. In this case, as the diagram indicates, the molecule has a central portion, represented by the large quarter-sized circle, and there is an outer electron of that molecule, represented by a small dime-sized circle. That electron is revolving clockwise. The negative elements in that right electron are also spinning in the same direction as the electron as a whole. The left area of the page shows a segment of a wire loop lying flat on the page. Inside the wire segment there are molecular nuclei, represented by the large circles, and electrons, represented by the small circles. When the magnet and wire have been brought into the relationship shown negative influences from the orbiting magnetic electron and from its spinning elements would push the electrons in the wire away and up the page. By convention it would be said that a current had been created flowing toward the bottom of the page.

The process just described is the means of which most electricity used today is generated. Some force, derived for example from steam or falling water, is employed to bring a magnet first toward and then away from a coil of wire. In this way an alternating current is produced in the wire. A current of this type is actually preferred for most uses since an alternating current is more efficient than a direct current for transmission of electricity over short distances. When a direct current is desired a set of switches, known as a commutator, can convert an alternating current in one wire into a direct current in another wire.

Initially when the magnet and the wire had been brought to rest with respect to each other the pressure on the electrons in the wire molecules nearest to the magnet would continue but, unless the wire happened to be in an extremely low temperature super-conducting state, the induced current in the wire would normally build up concentrations of electrons at points on the wire away from the segment next to the magnet, and these concentrations would increasingly oppose and soon halt the current in the wire. If the wire and the magnet were then pushed closer together another temporary flow of current would be induced. If the wire and the magnet were then pulled apart the concentrations of electrons in the wire would generate a temporary flow of current in the wire in the opposite direction.

When a wire and a nearby magnet are at rest with respect to each other and no induced electric current is flowing there is no significant net attraction or repulsion between the wire and the magnet. But when a magnet is being brought up to a wire, the moving electrons in the magnet are pushing electrons in the nearby segment of the wire against the concentrations of electrons built up further along the wire. As a result force is needed to push a magnet and wire together.

Conversely, when an electric current is flowing in a wire in a direction opposite to the electron spin in a magnet near the wire is that magnet is repelled away by the movement of the electrons in the wire. This interaction between a current-carrying wire and a magnet is the basis of electric motors. When an electric current is pushed through a wire the force projected by the wire can move a magnet mounted on the axle of a motor, and the spinning of the axle can run other equipment, such for example as a lathe.

There is also an effect when a wire which already has a current flowing in it is brought close to a material which is capable of being magnetized but is not already magnetized. If, for example, a wire were flat on this page, the flow of electrons in the wire were toward the top of the page, and the other material were nearby to the right of the wire the upward flow of the wire electrons would tend to push and pull the electrons of the other material into the configuration illustrated above on **Diagram 5**, thus making that material into a magnet with its north end pointing out from the page.

When a current is flowing in a wire another effect has also been observed. If there were nearby a magnet free to turn on a fixed axis parallel to the direction of flow of the wire's current the influences from the wire would twist the magnet until its north/south line was perpendicular to the current flow. If, for example, the wire were perpendicular to this page, the flow of electrons were into the page, the magnet were to the right of the wire, and the north end of the magnet was initially pointing to the top right corner of the page, then the negative influences emanating from the wire would push the outer electrons of the south end of the magnet more vigorously than the outer electrons of the north end of the magnet. As a result the north end of the magnet would be made to point directly up the page.

The word magnetism is now often applied both to the influences emanating from real magnets and to those emanating from electric currents, but in neither case is there anything involved other than straight-line Coulomb forces. What is unusual about influences from magnets, and not from influences from a current-carrying wire, is that these influences are likely to continue in a magnet when there is no electric current flowing from end to end of the magnet.

A current may also begin to flow in a wire when it is influenced by a stream of photons as discussed below in the section on Communication by Electric Radiation.

Light and Other Oscillating Electro-magnetic Radiation

A combination which revolves about an internal axis and has different concentrations of charge on different sides away from the axis radiates alternating net Coulomb influences. The net of these influences is strongest in the two directions exactly perpendicular to the axis of revolution. Revolving photons each composed of one positive positron and one negative electron are significant producers of this phenomenon although alternating oscillating radiation can be created in other ways as well. The frequency of alternating positive and negative influences arriving at some point from a photon would depend on the frequency of the photon's revolution and on the relative velocity of the photon and the point. When sufficient Coulomb influences alternating at frequencies between about 4.3×10^{14} and 7.5×10^{14} times per second arrive at a human eye, visible light is perceived. At the lower end of that frequency range red light is perceived. At the upper end violet light is perceived. In between in the range from the lower frequency to the higher frequency orange, yellow, green, and blue are perceived. When the eye receives a wide range of visible frequencies at the same time white is perceived. If no frequencies are received the eye sees black. The eye perceives light not because the eye is struck by photons, which it may be, but because parts of the eye react to the alternating positive and negative influences which came from photons and have reached the eye. In most cases these influences are from photons approaching the eye. When the source of light is far from an eye little of the significant Coulomb influence reaching the eye was projected by photons when those photons were near the source. Rather the predominant influences perceived by the eye are those which were projected by photons as those photons drew close to the eye. Photons striking a surface do, however, have small amounts of mass and momentum and can push some very light-weight objects a measurable distance and when concentrated by a magnifying glass can start a fire.[13]

When the time taken for radiation to travel between a source of radiation and a target is being reduced by motion of the source, the target, or both the frequency of the radiation arriving at the target appears to be increased, and vice versa, in the Doppler effect.

Frequencies in both directions beyond those which are perceived by the human eye can be detected by other instruments. In the ultra-violet direction are, among other types of radiation, x-rays and gamma rays. In the infra-red direction are, among others, thermal waves and radio waves. The range of frequencies which have been observed is wide, from frequencies of

10 per second to 10^{24} per second. That range can be expressed alternatively in terms of the distance a radiating force, say a positive one, would travel from a point before the next positive force in line reached that point, that is from about $10^{7.5}$ meters to $1/10^{15.5}$ meters.

From both the positive and negative components of a revolving bi-polar combination influence radiates out in straight lines in all directions to the extent the components don't block radiation from each other if those photons are not being subjected to Coulomb influences from other bodies while moving. If a point is stationary with respect to the source of such radiation, the intensity of the net radiation received at that point varies inversely with the square of the distance from the position of the source of the radiation to the point. Light radiation which left a source at the time of some event there would simply travel to the reception point at 300,000 km/sec. relative to that source. Relative to the target the radiation would move at 300,000 km/sec plus the velocity, if any, of the target directly toward the source of the radiation. The event at the source may well, however, have also caused revolving photons to travel from the source toward the reception point. Near the earth such photons travel at a velocity which is usually approximately 300,000 km/sec. relative to the earth. The ultimate level of perceived light radiation at the target would in that case be the sum of the radiation from photons at the action point and the later more intense, radiation from the moving photons. The total intensity of the light radiation received over any target area would, however, be reduced with distance from the action point both by the spreading of the radiation from each source and usually by thinning of the concentration of relevant moving photons at increasing distances from the action point. In most cases, however, the bulk of the radiation received at a particular point derives from the photons nearing that point. But at any point the radiation received from an un-concentrated stream of photons would include both radiation from photons moving directly toward the target and radiation from photons not moving directly toward the target.

It is not possible to concentrate the Coulomb emissions from a particular source but it is possible to concentrate the photons from a particular source into a narrow stream. In that case the intensity of total radiation received at a target is reduced less by distance from the action point. One way to accomplish this result is to bounce some of the photons leaving a concentrated source in a range of directions off of a concave reflector which sends those photons backward in parallel lines. Another way to arrange a narrow beam of photons is by use of a laser, which energizes photons inside a chamber and then lets them escape from the chamber only through a narrow hole pointed in a desired direction. By either method the radiation reaching a target is more concentrated than it would be if all the radiation reaching the target had emanated from widely dispersed photons. As a result of the influence of distance on radiation, however, even in the case of a narrow stream of photons the most influential radiation reaching a target would be that from photons which had reached points near the target.

The most intense radiation is received at a point if the axis of revolution of the approaching photons is perpendicular to the path of the photons to that point. And the intensity of radiation perceived is greatest at the point being approached by the stream of photons to that point.

A narrow beam of photons, from a laser for example, sends oscillating Coulomb influences sideways as well as forward but the sideways influences are so un-concentrated at any appreciable distance from the stream that they are normally not perceived. What is usually perceived along the path of a laser beam is the effect of the laser photons in lighting dust and vapor bits in the path.

Some of the aspects of light as described in the preceding paragraphs are illustrated in the **Diagrams 8 and 9.**

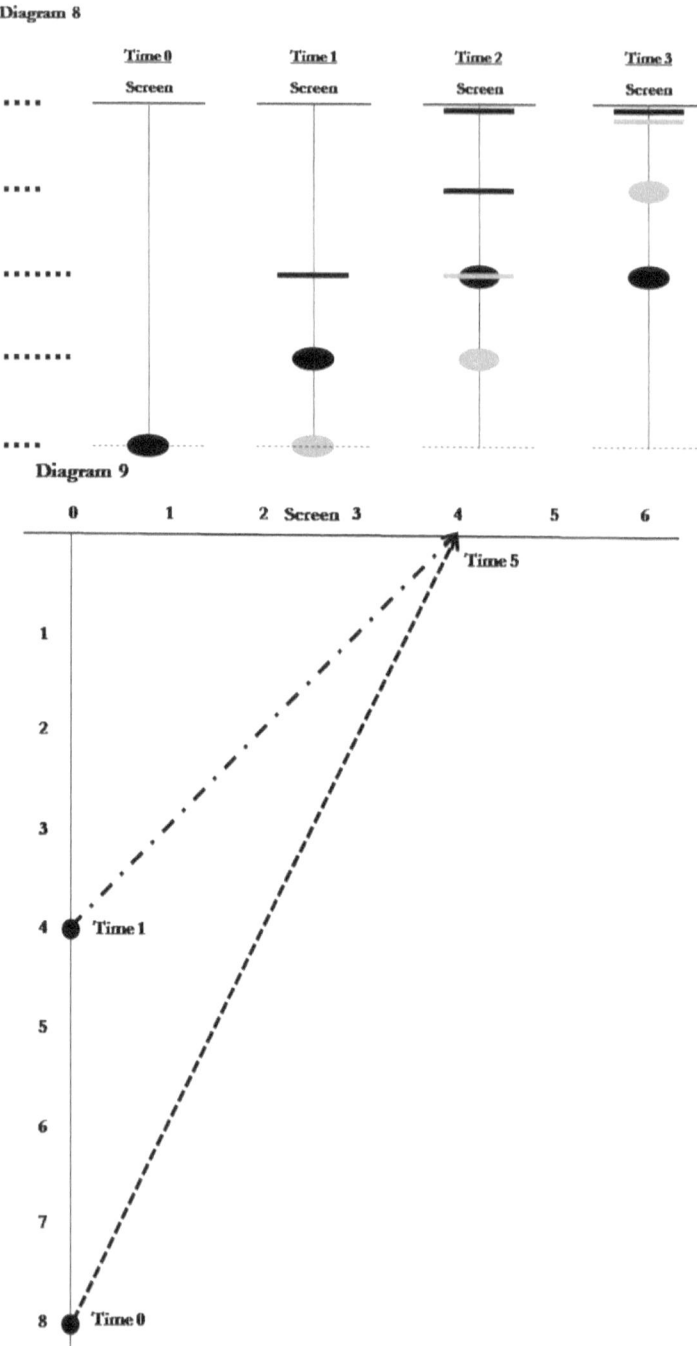

These diagrams display portions of the light radiations emanating from the movement of two photons assumed to be traveling up the page along the same path toward a point on a screen high on the pages. For simplicity the screen is assumed not to be moving with respect to the photons' point of origin. Each photon is assumed to be traveling at a velocity close to c and to be revolving about an axis perpendicular to the direction the photons are moving. The diagrams assume that in the illustrative cases the distance between consecutive photons is the distance that radiation would travel while a photon spun through 360 degrees about its axis. And at Times indicated by integral numbers it is assumed that the positive side of each photon is the side facing up the page. At those times each photon would radiate solely positive influence upward toward the point on the screen and a negative influence in the direction from which the photons came. In the other directions at those times net influences would be radiated. When a photon was half way between points at which it would radiate positive influences upward that photon would radiate net negative influences upward. The diagrams assume each bit of radiation moves at a velocity of c with respect to its source photon. As a result each bit of radiation moving in the same direction as its source photon moves at 2c with respect to the page, and radiation in other directions moves at a velocity between 2c and zero relative to the page depending on the direction of the radiation.

The first diagram shows at four consecutive times the position of two photons and the positive portions of the radiation from them moving directly ahead in the direction of the photons' movement. At Time 0 the photon shown as a black dot (for identification, not for identification of light color,) is at the origin point. At Time 1 that photon has moved a certain distance toward the screen and the original positive radiation from that photon (shown by a black horizontal line) has moved twice as far in the same direction. Also at Time 1 the next photon, shown as a grey dot, is originating radiation. By Time 3 the screen is experiencing reinforced same phase radiation from two different photons. At later times the screen would experience similar double radiation from other photons which left the point of origin at later times. And if the screen were further from the point of origin of the photons the screen would experience a sequence of reinforced radiations from more than two photons.

Diagram 9 shows two bits of radiation from two selected photons moving straight up toward a screen, but the bits of radiation shown are not the ones which went straight ahead from the photons but those which went from the photons toward the point K on the screen. The photons traveled upward at a velocity of one relative to the scale on the left. Starting at Time 1 the dashed-line radiation had traveled in its direction at a velocity of 1.415 as a result of the photon's velocity of one upward plus the radiation's velocity of one to the right away from its source photon. Starting at Time 0 the lower photon had sent the dotted-line radiation toward point K at a velocity of 1.79, somewhat greater than the velocity of the upper radiation. The two influences arrived at K both at Time 5 and both in the same phase. Although those influences were weaker than those arriving at D, the influences arriving at K did reinforce each other and, and if their frequencies were in the visible range, created a bright spot on the screen.

At other times at other points on the screen net influences would arrive in opposite phases and would then offset each other, so that the screen at those points would remain dark.

The phenomenon just described is a partial explanation why a narrow source of photons directed at a point such as D can create circles of alternating light and darkness around a well-lit target point. The circles of light would decrease in intensity the further they were from the target point since the radiation to the further points would have had further to travel from the emitting photons as compared to the radiation from the photons drawing near to the point D closest to the source. And the radiation arriving at K would be to some slight extent the net of both positive and negative elements whereas the radiation arriving at D was of a single type. The phenomenon just described, known as diffraction, partially explains why light could seem to bend to some extent around a corner located at the position of the blue photon at Time 0.

If there were two different streams of the same type of photons approaching two different points on the same screen from two points of origin, possibly from two slits in an opaque barrier, each stream could create an array of black and lighted spots along the screen. The electric influences arriving at various points, particularly those between the two points at which the photon streams had been aimed, could—depending on the locations of the screen and the sources of the streams—offset or reinforce each other.

There is also another process which can add to the diffraction from a stream of photons passing very close to a body or between two bodies which are close together. If, for example, a moving photon passes to the right of but very close to a revolving atom or molecule at the edge of a body, and at that time the negative side of the photon happens to be next to a positive side of the revolving atom or molecule, the path of the photon will be bent to the left. If the positive side of the passing photon is next to the positive side of the atom or molecule the path of the photon will be bent to the right. In this way, too, light can seem to some slight extent to bend around a corner, and play a role in observed interference or re-enforcement of light. Electrons, positrons, and larger bodies have also been observed sometimes to have their paths diverted in this way when those paths passed very close to other atoms or molecules, though the effects are more difficult to observe than with light. When photons or larger bodies are sent one at a time at separate times through slits or past a corner there is no interference or re-enforcement when the paths of the bodies are bent since the bodies do not arrive anywhere at the same time, but dark and light stripes can sometimes eventually be observed if a record of the impact points of the bodies on a screen is preserved.

It has also been noted that the path of a stream of photons can be bent when the stream passes near a massive body such as the sun, possibly both because the passing photons have been affected by gravitational attraction from the sun and because the photons have been nudged into revolving with a frequency which allows them to be attracted by alternating magnetic forces emanating from the massive body. As a result it has been calculated that the stream of photons arriving on the earth from a distant star do not always reach the earth from exactly the direction to the star from the earth.

When photons and Coulomb radiation reach a barrier, say a wall, all the photons and all the radiation may be stopped. In that case the wall is said to be totally opaque. In other cases some of the radiation and/or some of the photons may pass through the wall or some of the components of the wall may themselves be induced to emit photons. In that case the wall is said to be transparent to some degree. For photons the faster they are revolving, and the thinner the wall, the more likely is it that some light will be perceived beyond the wall. The degree of transparency of a wall also depends on the atomic and molecular structure of the material in the wall. In most cases the more orderly the structure the more likely that some photons and some radiation will find a clear path and pass through interstices in the material and come out the other side. The nearer the path of light approaching the surface of a body is to perpendicular to the surface the more likely is some of the light to pass through.

There is a possibility that photons reaching a wall will pass directly through without any collisions or significant interaction with the electrons and nuclei of the material in the wall. In other cases some of the photons may get through but only after following a zigzag path involving multiple collisions with combinations in the material in the wall. These collisions may cause no noticeable diminution of the speed of the photons in the various directions of the different legs of their travel through the wall, and the photons' speed after leaving the wall may be about the same as it was when they approached the wall. But the average speed of the photons in the direction directly through the wall would in this case appear to be less than the photons' speed before and after the wall. For this reason the speed of light passing through water, for example, appears to be less than the speed of the light through a vacuum.

When Coulomb radiation passes through a wall each part of that radiation which is not blocked by the wall continues in a straight line on the other side of the hole and does not spread out. However revolving photons continue to send out radiation in all directions after they pass through a wall. The total amount of radiation received from such photons is not the same in all directions from the location of the photon when it radiates. Since the photons are moving ahead the total amount of radiation received from such photons is greatest ahead in the direction of the photons' movement.

When photons emanating from a point source hit a surface made of some material at not too small an angle many of those photons enter the material in what is known as refraction. The angle below which no photons enter varies from material to material, but the materials which allow photons in tend to be materials with a very orderly structure. Once a photon enters such a material it has some chance of proceeding in a straight line through to the other side of the material without being bounced back or deflected by collisions with components of the material. The direction of the photon's passage through the material is usually not, however, a simple continuation of the photon's direction before it reached the material. When a photon with a certain frequency of spin passes from air or a vacuum into a denser but transparent material the photon retains its frequency but its direction of travel is usually bent into a new line closer to a perpendicular from the surface of the material. If that photon is not bounced back when it reaches the opposite edge of the material, and if the opposite edge is parallel to the edge first hit by the photon, then the photons direction will then be bent again back to its original direction but with some sideways shift from the original path as the photon leaves the material. The degree of bending when a photon enters and leaves a transparent material varies with the frequency of the photon's spin. A photon with a spin projecting blue light, for example, will bend more than a photon spinning more slowly and projecting red light. It is for this reason that a prism can be used to spread out the colors of an incoming white light. The bending of photons exiting a transparent medium such as water also explains why a fish under water will not actually be in the direction it seems to be when seen from a boat.

When a person is in a room with totally opaque walls without windows or a current source of light there may nonetheless still be some photons bouncing about in the room. But there would usually not be enough concentration of alternating radiation for him to detect it. The human eye probably needs at least six similar vibrations in rapid succession to perceive light. An exception would exist if there were present in the room some material which had recently had many orbiting electrons in its atoms joined by photons which pushed the resulting electron/photon combinations into further-out orbits from which those photons were only gradually being released. Such materials, called phosphorescent, could continue to give out visible light for several hours even in a dark room.

In the absence of phosphorescence, light could be created in the room in a number of other ways as well. One way would be to bring together in the room two, or more, chemicals which reacted so violently with each other as to create a lot of heat to activate more movement of photons, and possibly other types of combinations, present in or near the chemicals. Another way would be to apply so much heat to some part of the exterior of the walls of the room that the temperature increase was conducted through the wall and activated the interior of the wall to such an extent that that part of the interior of the wall gave off a visible concentration of alternating electrical influences. A third way would be to pass some alternating electric radiation through a hole punched in the wall. Even if the person in the room were not facing the hole he could possibly see light if the incoming radiation hit a point on the opposite wall at which the person were looking and induced enough alternating vibrations to send visible alternating electric influences back to the person. And revolving photons could be projected through the hole. Some of these photons could, when they reached the wall opposite the hole, send two different kinds of visible radiation back to the person. One type would be of the frequency of incoming photons bounced back to the person. These photons would radiate alternating electric forces which would allow the person to perceive the color produced by the frequency of the incoming photons. The other type of radiation could be of the frequency of the photons knocked off the wall to produce for the person the perception of the color of the wall.

Yet another way to produce visible light in the room would be to feed electric current into a wire in the room. If the combination of the strength of the current and the resistance of the wire to current flow were great enough the wire could be heated to the point of incandescence, heated enough that is that the wire would cast off a large number of photons which had been components of the wire. A wire into the room could also be used alternately to feed into some point of concentration in the room a large number of negatively-charged electrons and then to withdraw those electrons and more. That point of alternate concentration of net negative and than net positive charge would then radiate alternating electric influences which could have a frequency in the visible range.

Photons capable of radiating visible light could also be introduced into the room, even from a source a considerable distance away, through a flow of photons through a small plastic fiber optic tube.

Many sources of electric radiation produce simultaneously multiple frequencies. But whether a source is producing radiation of one or many frequencies that source may produce alternating influences from combinations revolving in different directions. When such multi-directional radiation passes through sheets of some types of transparent material only the radiation from spins in a certain direction is allowed to pass through. After passing through such material the radiation is said to be polarized. The radiation could then pass through another sheet of the same material oriented in the same direction as the first sheet, but if the second sheet were turned 90^0 from the orientation of the first sheet the radiation would be blocked from further travel.

When photons which emanated from a point source hit a flat surface, say a table top, some of those photons will bounce up from the surface. If the frequency of the radiation from those photons is in the visible range and there is a human eye, or other detector, some distance away above the table in a position where the photon source can't be seen directly the eye may still record light radiating from those photons which bounced, or reflected, in the direction of the eye. If the table surface is very smooth the photons redirected toward the eye will be primarily those which hit the table at the point where the angle between the table and the incoming photons is the same as the angle between the table and the outgoing reflected photons headed for the eye. Even if that particular point were cut out of the table, however, the eye would probably perceive some light since it would be unlikely that all incoming photons would hit exactly smooth points on the table. Some photons which hit at various points on the table would probably bounce toward the eye. But these photons would travel various different total distances from their source to the table and on to the eye.

If the photons all started out in the same phase of spin, that is with their, say, positive sides pointing in the same direction, later radiation from some of them reflected from different points on the table would arrive in the same phase while radiation from other photons reflected from other points would arrive out of phase. The intensity of the light perceived in this case would therefore be reduced as a result of interference between some of the positive and negative influences arriving at the eye at the same time.

It has also been observed that when a photon is reflected off a surface, such as a table top, the direction of its revolving is reversed. The collision not only changes the photon's direction of linear motion but also causes its spinning components to bounce back in the reverse spin direction. This change in direction of revolution would not make any change in the color of light perceived by an eye but could be detected by some detecting devices.

Electric Communication

Electric communication may now transmit more data than is transferred by physical movement of words and numbers written on paper. Communication by radiation in the form of flashes of light from shuttered lanterns has been employed for a very long time. And such light flashes could be read a long way off when they could be reflected off clouds. The efficiency of electric communication was much improved with the advent of telegraph wires. At first these wires just passed along an electric current in one direction, or didn't. Movement of much greater volume of information became possible when use was made of an alternating current moving back and forth in first a positive and then the negative direction. The frequency of the alternation of this current could be varied to send a message through what is called FM, frequency modulation, or the volume of the current could be varied to send a message through AM, amplitude modulation, and the efficiency of such transmission was further enhanced by adoption of the binary code by means of which letters, numbers, and even pictures could be transmitted by combinations of plus and minus bits. An oscillating electric current could be used to send successive positive and negative influences directly to a receiving station by wire, but the potential for communicating electrically was tremendously enhanced when techniques were developed for wireless transmission using frequencies other than those of light radiation.

For short distances electric communication can be achieved merely by varying Coulomb radiation directly from transmitter to receiver. For example, consider an antenna consisting of two wire segments some distance apart, one above the other not connected directly to each other but both connected to a source of electric charges. Suppose then that the source sends a positive charge to the upper wire and simultaneously a negative charge to the lower wire. The upper wire would then send positive influence in all directions including to the right and somewhat down where some distance away there was a small body with another positive electric charge. The incoming electric influence would tend to push the small body away and down. At the same time the negative influence traveling to the right and somewhat upwards from the negative lower wire would tend to pull the small body back toward the source and down. The net effect would be that the small body would not be pushed either away or back toward the source but would be pushed downwards. But if the source of charges quickly reversed the charges in the wires the small body would then be pushed upward. Manipulation of the charges and the timing of their alteration could then effect any desired pattern of up—and down oscillation of the small body, and if it were in a communications receiver the vibrations of the small body could be read as a message.

Messages in most wireless transmission are sent, however, not by transmission of Coulomb radiation directly from transmitter to receiver but by broadcasting revolving photons the radiation from which is delivers its message only when the photons approach the receiving station.

When the components of a wire without a current are activated by heating of the wire some photons are likely to be ejected. It is unlikely in this case that a predominant majority of these photons would be ejected while revolving in the same direction. But when a sizeable current flows through a wire, the bulk of the photons ejected in any particular direction are likely to be revolving in a common direction. When such photons approach another wire the revolving photons may push electrons in that wire in a particular direction to create an electric current. The direction of this induced current could then be reversed by an intentional reversal of the direction of the current in the source wire. In this way messages can be sent to be recorded, printed, displayed, or converted into sound at the receiving station.

Such transmission is most effective when the sending and receiving stations have line-of-sight relationships, but it was realized long ago that broadcast photons could in many circumstances be bounced off particle layers high in the atmosphere to send wireless messages for thousands of miles. It was also found that whenever a line of sight relationship was present—whether between two stations on the earth or between a station on earth and another in a space ship or satellite—the efficiency and security of communication could be increased by replacing broadcasting in all directions by focusing the outgoing photons into a narrow beam by reflecting them in a parabola-shaped dish and then by focusing the incoming photons at the receiving station toward a point with another dish.

Photons reflected by a parabolic dish on to a distant object can also be used in so-called radar devices to measure the distance to that object by measuring the total time taken for the photons to reach the object plus the time taken for photons reflected back from the object to be concentrated at the original source location.

Cosmology

Evidence gathered by astronomers suggests that all the matter in our universe was once compacted into a very small volume. There is no evidence as to what preceded that situation. And the evidence we do have does not rule out the possibility that there may be have been earlier universes or that there exist today other universes of which we are not aware.

It appears that on some occasion a very long time ago when our universe was compacted into a very small mass that mass exploded in what has been called the "Big Bang." That event occurred at least billions of years ago, but there is no reasonable way to specify precisely just how long ago since for part of the time since then the things we use to measure time, for example the motion of the earth around the sun and the behavior of cesium and carbon atoms, did not exist. In the immediate aftermath of the explosion everything material was apparently moving very rapidly and there were probably no large chunks of material. But over time smaller pieces of material coalesced into electrons, positrons, photons, protons, atoms, molecules, planets, stars, and possibly other things. It is now estimated that there are in our universe billions of stars since we observe photons coming from so many sources, but we don't really know how many stars there are for three reasons. First, there are so many of them. Second, there may be stars so far away that photons from them have not reached us. Even from the stars we have seen the light traveling to us must travel for many light years, that is many times the distance light travels in one year. And, third, the movements of some of the heavenly bodies which we can see suggest that there are out there a very large number of "dark" bodies which project forces but which do not send out newly-created light, or reflected light, which we can see. And there seem to be out there some bodies known as "black holes", which have such large masses compressed into small volumes that no photons passing near, them can escape from the "black hole's" immense gravitational attraction. Despite the probable existence of "dark bodies", photons created in the early stages after the "big bang" seem still be bouncing around and have been detected by earthly instruments.

It has never been possible to capture any dark matter in space and bring it back to the earth for examination. Perhaps it is just familiar material which just happens not to be projecting or reflecting photons toward us. The consensus view today is, however, that something on the order of a quarter of dark matter is not material of the types we are familiar with even though it appears to radiate gravitational attraction. And the consensus view is that there is also out there a lot of material even less like that we are familiar with. In a further degradation of the term energy this other material is generally referred to as dark energy. It is now projected that the dark material in our universe greatly exceeds the material of the type we are familiar with.

When a photon passes by a massive body in space without being captured the gravitational attraction of the massive body tends to increase the photon's velocity as the photon draws closer to the body and then tends to decrease the photon's velocity as the photon draws away. These velocity effects may, however, be swamped by the effects of collisions by the photons with other things, such as gas molecules which are often found in greater profusion near some large celestial bodies.

Many of the bodies in space are in long-lived orbits about other bodies, as in the case of the orbit of our moon about the earth and in the orbit of the earth and other planets about the sun. In general, however, the larger groups of bodies in space appear to be moving away from each other. Our universe is said to be expanding and the stars a long way from us are moving at a very high and increasing velocity away from us. Yet it is not clear why these velocities seem to be increasing so that the universe seems to be expanding at an increasing velocity. Perhaps way out there exists some type of inter-body influence, misleadingly called dark energy, which we have no experience with, or maybe the bodies we can "see" way out are being pulled into faster velocities by a massive number of other further-out bodies light from which is too weak for us to see.

Comments on the major differences between the model described in outline above and the currently conventional model are contained in the following section.

3

Footnotes on Differences in the Proposed Model from the Conventional Model

The paragraphs below are numbered according to superscript numbers shown at the ends of paragraphs in the presentation of the model in the previous section.

1. The basic nature of the universe posited in the opening paragraphs of the description of the proposed model differs from the current view of many physicists that there is no such thing as a material thing, that what we perceive as a thing is just a combination of perturbations in various all-pervasive fields in space, and that such perturbations can be made to appear or disappear any place at any time. In the proposed model space is treated as the emptiness between material things. It is not something which can be bent or warped. It does not originate forces. It is not something from which things can be created and into which things can disappear. Things can be combined or split apart but their component particles cannot be made to disappear.

2. The treatment of time in the proposed model differs from the Newtonian model in not assigning reality to something, that is independent time, which can not be measured. The treatment differs from that of the current Standard Model in not considering time to be a fourth dimension. A thing can understandably be considered to have more than three dimensions when the word dimension is simply used as a synonym for the word attributes. But the conventional physics model attempts to speak of a fourth dimension with some sort of spatial attributes as revealed, for example, by "warping" of space. The author believes, however, that a space dimension beyond three is a concept which the human mind can't really comprehend and doesn't need to try to. The conventional view is that when a clock is slowed it remains slowed when the force is no longer

applied. And the model differs from the assumption of Special Relativity that when the pace of a clock is slowed by a force applied to the clock the pace will usually automatically continue at the slower rate.

3. Apart from neutrinos and dark matter, the model treats only electrons and positrons as basic material particles. Current convention treats photons and quarks, the supposed components of protons and neutrons, as well as electrons and positrons as elementary particles. Quarks are however, a theoretical construct since it has never been possible to obtain any direct empirical data relating to an individual quark. The proposed model assumes that there are components and lots of space within bodies which are said to be composed of quarks but not that the structure of those components is known. For that reason no use is made of the concept of quarks.

4. In the proposed model there are not multiple types of action-at-a-distance influences, only the electro-magnetic Coulomb influences emanating from electrons, positrons, and combinations of them. In the model there are no separate types of gravitational, nuclear, magnetic, and electric influences. Influences from two or more separate charged bodies can combine to create enhanced Coulomb forces or can offset one another to result in reduced net Coulomb influences. In the proposed new model all Coulomb influences are transmitted on an action-at-a-distance basis. The more conventional view seems to be that all electro-magnetic influences are transmitted by what are called virtual photons. Since virtual photons have never been observed, and are assumed to have attributes which no observed material have ever been observed to have, it is not clear whether virtual photons are in agreement with the judgment once expressed by Einstein, who wrote "we have come to regard action-at-a-distance as a process impossible without the intervention of some intermediary medium." In the author's view any model must eventually rely on some action-at-a-distance without intermediary to hold together the left half and the right half of the smallest particle employed by the model. The proposed model treats Coulomb influences as always moving while they continue to exist. Although some texts speak of a field of force surrounding a charge, the author believes that usage can confuse the student since a "field" tends to convey an impression of something stationary and with unchanging dimensions.

5. The model considers that Coulomb influence always travels at the velocity of c with respect to its source but that the influence travels at different speeds, including possibly speeds in excess of c, with respect to bodies which are moving at different speeds relative to the source of the Coulomb influence. By contrast the conventional view accepts the

hypothesis of Einstein's Relativity that Coulomb influence always travels in a vacuum at a velocity of c with respect to any body as measured on that body whether that body is the source, the target, or a reference body for the influence and whatever motion the bodies may have. Special Relativity attempts to validate this hypothesis in a sense by "cooking the books" by assumptions that when a body had been accelerated relative to other bodies then thereafter the length of the first body would seem compressed and its clocks would have been made to run slow to different degrees as seen from the other bodies. This posited compression of bodies and variation in clock behavior has never actually been observed, however, and corroborative experience from velocity measurements with a target body traveling at a high speed relative to incoming Coulomb influence has also never been observed. Accordingly the hypotheses of Relativity mentioned above have not been incorporated in the proposed model. A more extensive analysis of the postulates of Special Relativity is presented in the next section.

Contrary to the currently conventional view that the velocity of reflected light is not affected by movement of the reflecting surface, the proposed new model assumes that reflected light leaves the reflecting surface at the same velocity relative to that surface as the light had in approaching that surface, and thus is affected by any motion of that surface This assumption is not contradicted by the famous Michelson/Morley experiment of 1887 which provided strong evidence that there does not exist an ether with which light, including reflected light, always travels at the same velocity. The results of that experiment were consistent with the possibility that light flashes approaching the experimental apparatus all traveled at the same velocity but that the velocity of the reflected lights within the apparatus had been affected by motion of the reflecting mirrors involved. This conclusion with respect to reflected light does conflict with a conclusion which Michelson presented in a separate lesser-known 1913 article, which the author finds unconvincing. The subject is elaborated in some detail in a separate later section.

6. There have been experiments which demonstrated that the accelerating effect on a charged body of a like-charged incoming Coulomb influence was reduced when the body was already moving in the direction of the influence. This effect is accepted by Relativity but is not unique to Relativity. The effect has been observed in both micro and macro experiments. A hose turned on a beach ball accelerates the ball less when the ball is already moving away from the nozzle. And the change in the velocity of a beach ball is greater when it is moving toward the nozzle.

The author is not aware of any experiment, for example with cyclotrons and linear accelerators, which measured the effect on acceleration of a very high velocity of motion by a target body in the direction opposite to that of incoming Coulomb force. Yet the proposed model assumes—based admittedly on a common sense hunch rather than experimental evidence at the micro level—that Coulomb influence will have greater effect on a body approaching a source of opposing influence and only the normal effect on a body moving perpendicularly to the direction of the Coulomb influence. The new model also assumes that the effect of Coulomb force is proportional to the velocity with which the influence reaches a target body. By contrast the conventional model assumes that the acceleration of a target body must be multiplied by the factor $(1 - V^2/c^2)^{1/2}$ whatever the direction of the velocity V of the target. This formulation, unlike that assumed in the proposed model, implies that the acceleration is affected by the motion of the target body to the same extent whether the motion of the target body is in the same direction or opposite direction of the incoming force, or sideways, and that a body's acceleration is reduced rather than increased when the target is moving toward the source of Coulomb influence.

7. In the conventional view, in addition to including the kinetic energy of a body as a whole in a particular direction relative to a reference body, the kinetic energy of the moving components of the body in various directions, and its potential in its position to have kinetic energy added, the total energy of a body is said to include also a so-called rest energy of mc^2, which would exist even if neither the body as a whole nor any of its components were in motion. In the new model proposed above there is no such thing as rest energy. And unlike conventional textbooks the presentation of the new model above does not attempt to explain the rules of physics primarily in terms of energy. The author believes this reliance on energy is un-necessary and actually harmful in teaching. Some textbooks state that a body may also have heat energy but that is just another word for the kinetic energy of the moving parts of the body. Some texts state that a body may have chemical energy, but that is just a form of potential energy, as illustrated in the main text.

8. In contrast to the proposed new model, in which the photon is composed of two particles each of which has both charge and mass, the conventional view is that a photon at rest would have no charge and no rest mass and yet that a moving photon could convey charge, momentum, and energy. Puzzling to the author, moreover, is the fact that in the conventional view the energy of a photon is measured by its angular velocity with no account

taken of its linear velocity. It seems to me that some abuse of language is being committed by this conventional view. It is, of course, difficult to measure the mass of a rapidly-revolving photon, but since a photon is sometimes converted into a positron and an electron, particles which clearly do have mass, it seems reasonable to the author to assume that the photon is composed of a positron and an electron, as is done in the new model. In contrast to the new model's assumption that photons usually leave their source at a velocity of c relative to that source and arrive at their target at a velocity relative to the target which depends on the relative motion of the source and target, the conventional view is that photons would always be measured to have a velocity of c both on the source and on the target.

9. The conventional view is that the proton is composed of a down quark, with a charge of minus one third, and two up quarks, each with a charge of plus two-thirds. This convention does provide the proton with the observed net charge, but since individual quarks have never been observed I see no advantage in considering protons to be composed of quarks rather positrons and electrons. The conventional view is also that the components of a proton are not held together by simple Coulomb influence but rather by some separate form of strong force. The author believes that the electron and positron components of a proton can be so structured that simple Coulomb influences can hold the parts of the proton together.

10. The conventional view is that the orbital radius of an electron to which a photon has become attached is explained, not by the considerations described in the text above, but by the assumptions that there are some specific basic "quantizing" rules in physics and that the photons involved in the process are created out of nowhere and later disappear into nowhere. It should be possible eventually to test the hypothesis of the new model by actual measurements of masses, velocities, and orbital radii involved.

11. When the north end of one magnet is below the south end of another magnet then in the conventional view magnetic attraction between the magnets is somehow caused primarily by the electrons at those ends of the magnets spinning in the same direction about their internal axes not actually but in some theoretical sense The author is skeptical of the apparent assumption that with the help of Special Relativity two electrons spinning in the same direction and traveling side-by-side in the same direction would ever attract each other, particularly if the spins were not real. The proposed model considers magnetism to be the result of the projection of positive and negative Coulomb influences in particular directions as electrons in magnetic substances rotate around their nuclei

and the projection of Coulomb influences from actually spinning segments of electrons in the magnets.

12. Unlike the proposed new model the conventional view attempts to explain the attraction of parallel wires with currents flowing in the same direction by assuming a la Relativity that the movement of the electrons in a wire makes their spacing seem contracted from the viewpoint of the other wire. The charges in that contracted wire are somehow supposed then to result in a net attractive force, a contention the author finds unpersuasive. Also since the writings of James Clerk Maxwell it has been the conventional practice to teach that current moving through a wire emanates two types of influence: one magnetic and one electric, which point in different directions. The author finds that usage misleading since there is only one type of influence emanating from a current-carrying wire though charged bodies and magnets do interact differently to that influence.

13. Conventional presentations regarding light do not make a clear distinction between the roles of moving photons and Coulomb radiation. The student is just told that sometimes light should be considered a wave and at other times should be considered a particle. And there are some physicists who believe that photons in the form of combinations having mass and charge do not exist and that all effects ascribed to photons can be explained by Coulomb radiation. Yet it appears that electrons and positrons, which do seem to have mass and charge, sometime combine to form photons, which are considered to be real things.

The currently conventional view, based on Einstein's Relativity, is that every light flash in a vacuum should be measured as traveling at the velocity of c with respect to every body regardless of any motion of that body with respect to other bodies. One corollary of this view is that the velocity of a flash of reflected light is not affected by the movement, if any, of the reflecting surface. By contrast the proposed new model considers that the velocity of a photon is influenced by the movement of a surface from which that photon has reflected, that, while most photons encountered near the earth originated in the sun with a velocity of c with respect to the sun, those photons when encountered near the earth have been reflected from surfaces on the earth and have acquitted a velocity of c with respect to the earth. Some background information with respect to the model's treatment of light is presented in the penultimate section of this booklet in a section entitled "Comments on Two Light Velocity Experiments".

4

The Proposed New Model in Summary

1. Apart from some extremely small things called neutrinos, and possibly some very large things far out in space, called dark material, about which little is known, all the things we encounter are combinations of electrons and positrons. The smallest stable combinations are the photon, composed of one electron and one positron, and a proton, composed of approximately 918 positrons and 917 electrons. There is no known explanation why there are no stable combinations intermediate in size between these two combinations. Many larger combinations are stable.

2. Each electron and positron exhibits the same resistance to being moved and is said to have the same mass and inertia. The mass of a combination is the sum of the masses of its positrons and electrons.

3. A massless action-at-a-distance influence, known as a positive Coulomb influence, emanates from each positron, which is said to have a positive charge, and a massless negative Coulomb influence emanates from each electron, which is said to have a negative charge. The charge of a positron has the same magnitude as the charge of an electron but the two types of charges have opposite effects on other bodies. There are no other types of action-at-a-distance influence. The net charge of a combination is the algebraic sum of the charges of its constituent positrons and electrons.

4. Coulomb influence is always emitted in a vacuum at a velocity of c relative to its source. The velocity of that influence with respect to a target depends on the velocity of the target relative to the source. Positrons, electrons, and combinations can move at various velocities but most photons encountered near the earth originally left the sun with a velocity of c with respect to the sun. After reflecting from surfaces on the earth those photons usually have a velocity of approximately c with respect to

the earth. The velocity of a reflected photon is influenced by the velocity of the reflecting surface.

6. Coulomb influence from a charged combination accelerates a like-charged combination away from the source of the influence to an extent which is proportional to the product of the net charges of the two bodies, proportional to the velocity of the influence with respect to the target body, inversely proportional to the mass of the target body, and inversely proportional to the square of the time taken for the influence to reach the target body; and Coulomb influence from a charged combination accelerates an oppositely-charged combination toward the source of the influence to an extent which is proportional to the same factors.

7. Gravity is not a separate form of action-at-a-distance influence. Gravitational attraction exists between large groups of combinations when passage between the groups of some of the negative influences from outer electrons in the molecules of the combinations is blocked by concentrations of electrons and positrons in the cores of those molecules.

8. Magnetism is the sequence of alternative Coulomb influences emanating from a group of molecules spinning at the same pace in the same direction about parallel axes, with the result that two magnets in a line with their north ends pointing in the same direction attract each other and electric current is induced when a magnet is moved toward a wire which is perpendicular to the north/south line of the magnet.

10. Light is perceived when alternating positive and negative Coulomb influences emanate from photons spinning around parallel axes in a particular range of frequencies with a timing such that the influences from different photons arrive at the eye, or other detecting device, in a manner to reinforce each other.

11. Time is not a fourth dimension. Measurement of a length of time is just a comparison of the extent of movement of one material combination with the extent of movement of another material combination.

5

An Illustration of the Hypotheses and Implications of Special Relativity

In his famous first published paper on Special Relativity in 1905, Einstein stated that his conclusions were derived from two basic hypotheses. He later added that these hypotheses are only applicable to domains "in which no gravitational field exists." His *first basic hypothesis* was "The laws by which the states of physical systems undergo change are not affected whether these changes of state be referred to one or the other of two systems of co-ordinates in uniform translatory motion," that is in motion at a constant velocity in a straight line. He later provided a more general version of the first hypothesis writing that all coordinate systems in which "readings" "observed simultaneously on adjacent clocks (in space) differ from each other by an indefinitely small amount" "are essentially equivalent for the formulation of the general laws of nature." He sometimes referred to such a system of coordinates as "a stationary system of coordinates." Under this more general formulation the laws of physics are the same in two such systems of coordinates even if their relative motion is not uniformly translatory with respect to each other, that is even if one of the systems of coordinates is accelerating with respect to the other.

His *second basic hypothesis* was "Any ray of light moves in the "stationary" system of co-ordinates with the determined velocity c, whether the ray be emitted by a stationary or by a moving body." This hypothesis has been interpreted to mean that any ray of light would be measured to have a velocity of c when measured in a stationary system of coordinates whether that velocity was measured on the body which was the source of the light, on the body which was the target of the light, or on any other reference frame and whether the body with the stationary system of coordinates was moving or not. The statement requires only that the two points between which the speed is being measured must be stationary with respect to each other at the time of measurement. Any two points stationary with respect to each other could provide the system of coordinates for measuring the speed even if those two points were moving with respect to other points.

Although the second basic hypothesis was stated in terms of light the hypothesis was considered to apply as well to all forms of electro-magnetic radiation.

In addition to those two basic hypotheses there are four other hypotheses implicit in Special Relativity, as discussed below. The hypotheses are illustrated by an example which shows what are, according to Special Relativity, the effects of a force which acts on one body and not on another upon measurements taken on those bodies of the velocities and lengths of the bodies and of the readings of clocks attached to them.

Suppose that there were at some time two space ships, as depicted in the table below, parallel to each other and at rest with respect to each other out in the near-vacuum of space in a region with no discernible net gravitational or electro/magnetic influences other than one electric force specifically described below.

Table **I**

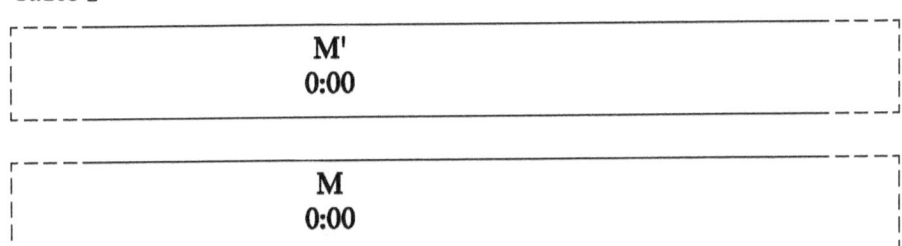

Each rectangle in the table represents a space ship with its front considered to be to the right. Any two points both on either one of the ships are considered to provide a stationary system of coordinates. Each ship has a mass of 100,000 kilograms (kg). The upper ship has a net positive electric charge distributed among its parts in proportion to their masses. The lower ship has no net electric charge. Each ship has clocks at many points along its length and next to each clock a camera recording continuously the reading of that clock, the reading of the adjacent clock, if any, on the other body, and the position of the clocks on the bodies relative to distance markers on those bodies. The colons in the table indicate the positions of selected clocks. The number to the right of a colon indicates the reading of the clock at that point in nanoseconds (ns). The number to the left indicates the distance in meters (m) to that point as measured on that body from its midpoint, which on the upper ship is marked M' and on the lower ship is marked M.

Pictures taken of the data at any specific pair of adjacent points on the ships would always be the same whether the pictures were taken by a camera on the lower ship or by a camera on the upper ship.

When the cameras recorded the numbers shown in the table above all the clocks on the ships were synchronized and read zero. Starting then and continuing for 50 ns on the clocks of the lower ship a constant positive electric force of 800 kg m/ns/ns was applied to the positively-charged upper ship pushing it to the right. The force was applied to all parts of the upper ship in proportion to the masses of those parts. The magnitude of the force was measured with respect to the lower ship. The force in this case had its origin in the inter-action between the positive charges on the upper ship and the positive charge on a body other than those in the table and did not affect the lower ship. Such a constant assumed force is convenient for an illustrative example but would be very difficult to arrange in practice. The distances from the other body to the upper ship at the times of radiation of the other body's influence would probably be constantly changing in practice. The distances the radiation would have to travel after emission to reach the upper ship would also be changing as that ship's velocity varied.

As mentioned above in the discussion of the model, Special Relativity hypothesizes that all light or electrical influence arriving at a body, such as the upper ship, will always be measured on that body to have the same velocity, regardless of the relative motion of that body with respect to other bodies, so long as the velocity is measured by use of two points on that body constituting a stationary system. Nevertheless, according to Special Relativity, incoming influences measured on the lower ship to be equal would be seen to have different effects on the upper ship when the upper ship is moving at different velocities with respect to the lower ship as those effects are measured on the lower ship. According to Special Relativity, a force F of a certain size as measured on the lower ship would somehow give acceleration A to the upper ship to a lesser extent relative to the lower ship when the upper ship was in motion away from the source of the incoming positive influence with a greater velocity V with respect to the lower ship than when the upper ship had lesser velocity. Under the rules of Special Relativity the effect of a body's velocity in reducing the accelerating effect of an incoming influence is not simply proportional to the ratio of the body's velocity to the velocity of the incoming influence. Rather, according to *one implicit hypothesis* the effect can be expressed precisely in either of the following formulations, in which W represents the mass of the body receiving a force F and c has a value of approximately 300,000 km/sec, which shall here be considered to .3 m/ns:

$$F = W \times A/(1-V^2/c^2)^{.5} \quad \text{or} \quad A = (1-V^2/c^2)^{.5}F/W$$

As a result when a constant force is applied to an accelerating body over equal increments of time smaller and smaller increases in velocity are hypothesized to result as the body's velocity increases. If a body's velocity approached the value of c then according to the formulas its acceleration would approach zero and the resulting velocity could never exceed c.

With the constant force and the values for c and mass assumed above the equation for acceleration of the upper ship as measured on the lower ship becomes:

$$A = (1 - V^2/.3^2)^{.5} \times 800/100,000 = .008(1 - V^2/.3^2)^{.5}$$

When the acceleration of the upper ship relative to the lower ship is considered to be a function f" of time T the equation for acceleration can be restated as:

$$f''(T) = .008(1 - (f'(T))^2/.3^2)^{.5}$$

By integrating from this formula it is possible to calculate that in 50 ns the movement of the upper ship as measured on the lower ship was:

$$f(T) = 11.25(1 - \cos[2T/75]) = 8.6035 \text{ m.}$$

This movement is less than the 10 m the movement would have been if the velocity of the ship had not reduced the accelerating effect of the incoming force.

At the end of the 50 nanoseconds the velocity of the upper ship would be:

$$f'(T) = .3\sin(2T/75) = .2916 \text{ m/ns}$$

At that time the velocity would have been .4 m/ns in the absence of the effect of the velocity in reducing acceleration.

Over the period of acceleration the mass of the upper ship did not change but its momentum and kinetic energy were increased.

According to a *second implicit hypothesis* of Special Relativity a body, in this case the upper ship, which had been at rest with its clocks synchronized with the clocks on a reference body, in this case the lower ship, would have the pace of time-keeping advance of its clocks, as measured on the lower ship, lowered relative to the pace of timekeeping advance of clocks on the reference lower ship when a force had accelerated the clocks on the upper ship into motion relative to the clocks on the lower ship. At each velocity of the upper ship's midpoint clock relative to the reference body the pace of timekeeping advance of that clock would be measured at $(1-V^2/c^2)^{.5}$ times the pace of timekeeping advance of the clocks on the reference body, considered here to be one ns per ns. Accordingly the pace of timekeeping advance of the upper ship midpoint clock after 50 nanoseconds would be:

$$\{1 - (.2916)^2/.3^2\}^{.5} = .2353 \text{ ns/ns.}$$

By integration it can also be determined that over the full 50 nanoseconds of the period of acceleration on the lower ship's clocks the total timekeeping advance of the upper ship's midpoint clock would be 36.4476 ns. On the basis of the calculations just described the calculated position and reading of the upper ship midpoint clock relative to the lower ship at the end of the 50 ns period on the lower ship is shown in the following diagram. At that time the upper ship's clock was no longer synchronized with the clocks on the lower ship.

Table II

	M' 0:36.45	

	M 0:50	8.60:50

After the end of the period of acceleration the velocity of the upper ship midpoint clock as measured from the lower ship would continue to be the .2916 m/ns achieved at the end of the period, and the pace of timekeeping advance of the upper ship clock as measured on the lower ship would continue to be .2353 of that of the lower ship clocks. The next table shows then what would be the positions of the upper midpoint clock and the adjacent lower ship clocks an illustrative 28.6783 ns after the end of the period of acceleration as measured on the ship.

Table III

	M'	
	0:43.20	

M		
0:78.68	8.60:78.68	16.97:78.68

The upper midpoint clock had moved a further 8.3626 m to the right relative to the lower ship, and the upper clock reading had advanced a further 6.7480 ns while the ship's clocks advanced 28.6783 ns.

The two tables above illustrate a result of Special Relativity often referred to as the **Twin Paradox**. If when the ships had been at rest with respect to each other there had been on board the lower ship at its midpoint a twin whose aging process proceeded like clockwork and at the midpoint of the upper ship the other twin with the same aging process then the lower twin would observe—when later studying pictures taken by the cameras on the lower ship—that over the acceleration period the upper twin had aged only about three quarters as much as he or she had. During periods after the end of acceleration the upper twin would appear to have aged only about as quarter as much as the lower twin.

Since in the case just discussed the force was applied to all parts of the upper ship in proportion to the mass of the parts, the force would have caused no physical stretching or contraction of the ship as measured on that ship, but according to a *third implicit hypothesis* of Special Relativity a body which is accelerated relative to a second body would have all segments of its length, as measured on that second body at the end of the period of acceleration, contracted to $(1-V^2/c^2)^{.5}$ times the previous rest length of that segment, when V is the velocity attained by all parts of the body relative to the second body at the end of the acceleration period. The apparent contraction would be toward the center of gravity of the contracting body. The following table then shows what the distances from the upper ship midpoint to two other illustrative upper ship points, A' and B', were after

50 ns on the lower ship clocks from two different viewpoints, that is as measured on the upper ship and as measured on the lower ship at the end of the period of acceleration. The selected illustrative upper ship points A' and B' had both been 72.1041 m from the upper ship midpoint as measured from both ships before the acceleration. After the acceleration the selected points were still 72.1041 m from the upper ship midpoint as measured on the upper ship but only 16.9661 m from the point on the lower ship adjacent to the upper ship midpoint as measured from the lower ship.

Table IV

A'	M'	B'
-72.10:	0:36.45	72.10:

		M		
	-8.36:50	0:50	8.60:50	25.57:50

During the acceleration period as measured on the lower ship the A' point on the upper ship moved 63.7415 m, from -72.1041 m to -8.3626 m, that is faster and further than did the upper ship midpoint, which moved only from 0 to 8.6035 m. The B' point, as measured on the lower ship, actually moved in the direction opposite to the upper midpoint's movement, that is from 72.1041 back to 25.5696 m.

The following diagram shows the different distance measurements 28.6783 ns after the end of the acceleration period. During this period all points of the upper ship moved 8.3620 m to the right relative to the lower ship at .2918 m/ns as measured on the lower ship and experienced no further acceleration. During that time there was no further contraction of the upper ship as measured from the lower ship.

Table V

A'	M'	B'
-72.10:	0:43.20	72.10:

	M	
0:78.68	16.97:78.68	33.92:78.68

The two tables just above did not show the readings of the clocks at points A' and B' on the upper ship. To calculate these readings it is necessary to take into account the fact that under the rules of Special Relativity the apparent advance of accelerated clocks other than those at the center of gravity of a body, as seen from an un-accelerated body, is not governed by the hypothesis, used above for the upper ship midpoint clock, that the advance of an accelerated clock is determined solely by the velocity experienced during the period of acceleration. For clocks not at the center of gravity on a body accelerated from a rest position a *fourth implicit hypothesis* of Special Relativity provides that at the end of the period of acceleration the reading of a clock which is ahead of the center of gravity will be behind the clock at the center of gravity by VxL/c^2 where V is the velocity of the body at the end of the acceleration and L was the distance between the two clocks before the acceleration began, all as measured on the reference body. By the same hypothesis the reading of a clock behind the center of gravity will be ahead of the center of gravity clock by VxL/c^2. The readings of the ship clocks under this hypothesis at the end of the 50 ns acceleration period on the lower ship's clocks are shown in **Table VI** below. While the upper ship midpoint clock moved forward by 36.4476 ns during the acceleration period, as measured from the lower ship, the point A' clock, which had presented a zero reading at the same time the other clocks did, moved forward by more, that is by 270.0649 ns, to a reading 233.6173 ns ahead of the clock at the upper ship midpoint as measured on the lower ship. During the acceleration the forward upper clock ended up 233.6173 behind the upper middle clock at -197.1697. The reading of the upper ship forward clock as seen from the lower ship actually moved backward, therefore, during the acceleration period.

Table VI

A'	M'	B'
-72.10:270.05	0:36.45	72.10:-197.17

		M	
-8.36:50	0:50	8.60:50	25.57:50

(As revealed by the table above, if a twin were at the A' point on the upper ship, rather than at the upper midpoint as discussed earlier, that twin would have aged more than a twin at the lower midpoint over the acceleration period, but if the upper ship continued to move at .2916 m/ns after the end of acceleration the upper twin would quickly fall behind the lower one in age.)

As also shown above, the acceleration period, which was 50 ns on all the lower clocks, appeared to have durations of various lengths of time at different points on the upper ship as measured by the clocks at those points

(If the cameras on the two ships had photographed, not only the upper midpoint clock at 36.45, but also in the same picture the clocks on either side of the midpoint clock then according to the Relativity hypotheses the photograph would have revealed that the clock to the right had advanced beyond 36.45 and that the clock to the left had not advanced to 36.45. In that event an observer on the upper ship, as well as an observer on the lower ship, could see evidence of the differential effects of the hypotheses on the clocks of the upper ship.)

The next table shows all the clocks after a further 28.6783 ns on the lower ship clocks. During that period all the upper clocks appeared to move 6.7480 ns at the same pace.

Table VII

A'	M'	B'
-72.10:276.80	0:43.20	72.10:-190.42

	M	
0:78.68	16.97:78.68	33.92:78.68

If a single point force, rather than the distributed force assumed above, had been applied to the upper ship, say just to the selected point A', then it would have taken time for the effect of the force to travel by electro-magnetic influence from particle to particle from point A' to other points in the ship. The A' point of the ship would therefore have moved sooner than a point ahead, thereby compressing the part of the ship forward of point A'. And the rear part of the ship behind the A' point would have been stretched. This compression and stretching would last as long as the ship was being accelerated by the force. As a result the positions shown in **Table VI** above for the various points 50 ns after the distributed force had been begun to be applied would not be applicable at the exact end of a 50 ns acceleration by a single force applied to a point on the ship.

But if the ships were constructed out of material which made them totally resilient in the sense that when one was held in place but compressed or stretched there developed within the body a countervailing force which would, sometime after the compression and stretching force had ended, return the body to its original shape and size, if no other forces were then affecting the body's shape and size, then the compressing and stretching time effects in the upper ship would be quickly offset after the end of the acceleration. **Table VII** above of positions after a total of 78.6783 ns on the lower ship could then properly represent the configuration of the ships as seen from the lower ship no matter where the force had been applied for 50 ns.

The second basic hypothesis of Special Relativity indicates that a flash of light, or of other electro-magnetic radiation, passing between two points of any body using a stationary reference frame should be measured on that body to have a speed of c regardless of the movement of that body with respect to other bodies and regardless of the source of the flash. With the changes in the positions of points on the upper ship and in the readings of the upper ship clocks, all as seen from the lower ship, as illustrated above in accordance with the hypotheses of Special Relativity, it can be calculated that a light flash would be measured at the same speed on the upper ship as on the lower ship even though the ships were moving with respect to each other. For example, consider a light flash from dead ahead of the accelerated ship, that is from the right. According to the second basic hypothesis it would make no difference whether the source of the flash were at rest or moving with respect to the space ships. Suppose the flash reached the upper ship midpoint M' and the adjacent 8.6035 m point on the lower ship at the end of the period of acceleration for all clocks when the clock at the upper point read 36.4476ns and the adjacent lower clock read 50 ns, all as shown in **Table VI** above and repeated in part in **Table VIII** below.

Table VIII

		M'
-72.10:270.05		0:36.45

		M
-8.36:50	0:50	8.60:50

On the assumption that the flash was moving at .3 m/ns relative to the lower ship that flash would travel 8.6035 m from the lower 8.6035 m point and reach the midpoint on the lower ship 28.6783 ns later when the clock there read 78.6783 ns, as shown in **Table IX** below. When the flash reached that point on the lower ship the -72.1186 m point on the upper ship would have moved, as measured from the lower ship with a velocity of .2916 m/ns for 28.6783 ns for 8.3626 m to a position adjacent to the lower ship midpoint. And the left clock on the upper ship would, like the upper midpoint clock, have advanced by 6.7480 ns to 276.8129 ns. The light flash would thus reach the lower ship midpoint and the then adjacent -72.1041 upper ship point on the same occasion in the light's travel. During its travel from the lower ship point adjacent to the upper ship midpoint to the lower ship midpoint the flash traveled 8.6035 m in 28.6783 ns as measured on the lower ship, a measured light velocity of .3 m/ns. During the same segment of its travel the flash traveled, according to measurements on the upper ship, 72.1041 m from a point with a clock reading of 36.45 ns to a point with a clock reading of 276.81 a duration of 240.36 ns, therefore also at a measured light velocity of .3 m/ns.

Table IX

A'	M'
-72.10:276.80	0:43.20

M
0:78.67

Similar measurements on both ships of light approaching the ships from other directions would, according to the hypotheses of Special Relativity, also confirm the constancy of light speed measurements.

If the force described above had pushed the upper ship in the opposite direction relative to the lower ship, measurements taken on the lower ship would have recorded the same change in velocity magnitude, the same length contraction, and the same type of re-adjustment of the clocks of the upper ship as detailed above. After the acceleration the clock at the midpoint of the upper ship would have been advancing at the same rate as illustrated earlier, a rate slower to the same degree relative to the clock on the lower ship.

If after the 50 ns of acceleration to the right described above for the upper ship it had been slowed down to a position at rest next to a position, not the original one, on the lower ship then the length of the upper ship and the pace of its clocks would be returned to their original magnitudes as measured on the lower ship. All the upper ship clocks would then appear to the lower ship to have the same reading as each other, but that reading would remain behind the reading of the lower ship clocks.

If the upper ship's movement to the right were brought to a halt, the ship were then accelerated back to the left and decelerated to a halt in its original position next to the lower ship the upper ship's length and pace of clock advance would have returned to their original magnitudes, but the return journey would have put the readings of the upper ship clocks even further behind the clocks of the lower ship.

If at some time prior to the zero time of the original example above some force other than the force described above had pushed both ships into a common velocity to the left relative to another body Q relative to which the ships had been at rest sometime earlier in the past with synchronized clocks then at the zero time above the clocks on both ships could still have been synchronized but all those clocks would, in the eyes of observers on Q, be advancing more slowly than clocks on Q. If thereafter at some time later than the times displayed above another small force toward the right were applied to the upper ship a contradiction would seem to arise between the rules of Relativity as modified after 1905 to provide that the laws of physics are the same even on accelerating bodies. When the upper ship was given the small push by the last-mentioned force that ship would acquire increased velocity and therefore a slower rate of clock advance as measured on the lower ship. Since Q and the lower ship were not affected by the last-mentioned force the slowing of the upper ship's clocks relative to the lower ship would also seem to imply a slowing of the upper ship's clocks relative to clocks on Q. On Q, however, the last-mentioned force would seem to have reduced the leftward velocity of the upper ship so that observers on Q would be expected to see the timekeeping of the clocks on the upper ship speed up rather than slow-down. And what would be the situation with respect to some other body P relative to which the upper and lower ship and Q had been at rest at some earlier time before some force pushed them apart. There seems to be a conflict between the rule that accelerated motion results in the appearance of slower clocks on the accelerated body as seen from a body not experiencing the accelerating force and the rule that slower motion of a body as the result of a decelerating force should provide the appearance of faster advance of that body's clocks as seen from a body not experiencing the decelerating force.

All of the diagrams above presented the numbers as seen when all the lower ship clocks had a common reading. The diagrams below will demonstrate the creation an illustration in which all the numbers are presented with common readings on the upper clock—still in the case where the force was applied to the upper ship.

On Table X below are shown three pairs of points which were side-by-side on the ships at Time 0 on the clocks of both ships just before acceleration of the upper ship began

Table X

A'	D'	H'
-72.1:0	-70.14:0	-68.17:17

A	D	H
-72.1:0	-70.14:0	-69.17:1

As shown in **Table XI** below the point **A'** had moved to the right to be beside the lower ship point P 50 ns later on the lower ship clocks at the end of the acceleration period on those clocks. At that time the reading of the upper ship clock at **A'** had changed to 270.05

Table XI

A'
-72.1:270.05

P	M	
-8.36:50	0:50	8.6:50

A further 27.06 ns later on the lower ship's clocks, as shown on **Table XII** below, point **O'** on the upper ship had moved to the right to a position adjacent to point **M** on the lower ship and the reading of the clock at **O'** had changed to 270.05, the same time displayed by the clock at upper ship point **A'**.

Table XII

D'
-70.14:270.05

M
0:77.06

And yet another 27.06 ns later on the lower ship clocks **Table XIII** below shows that the upper ship point **H'** has moved adjacent to the Q point on the lower ship ands the upper ship clock at point **H'** has moved to the same 270.05 ns value that **A'** and **D'** had in the previous Tables.

Table XIII

H'
-68.17:270.05

Q
8.36:104.1

On the next Table **XIV** all the data from the previous three Tables is presented with a common reading of 270.05 ns on all the upper ship clocks as provided by the previous three Tables for the three selected upper ship points. The lower clocks in **Table XIV** no longer show a common reading. The distance as measured on the upper ship between the lower ship point **P** and the zero midpoint **M** on the lower ship is shown to be 1.96 m whereas the separation of those two points was 8.36 m as measured on both ships before the acceleration began. And the distance as measured on the upper ship between the lower ship midpoint and the lower ship point **Q** is shown as 1.97 m whereas the separation was 8.36 m before the acceleration began. Over the time period on the upper ship clocks from the start of acceleration to readings of 270.05 ns, as the lower ship moved to the left as seen from the upper ship, the lower ship appeared to contract as measured on the upper ship, whereas during the acceleration period distances on the upper ship appeared to contract as measured on the lower ship. The time periods are not exactly comparable, however. The time period on the upper ship from zero to 270.05 ns includes for some upper ship points segments of time after those points stopped accelerating.

Table XIV

A'	D'	H'
-72.1:270.05	-70.14:270.05	-68.17: 270.05

P	M	Q
-8.36:50	0:77.06	8.36:104.1

The tables above also make clear that while the upper ship appeared to move to the left as seen from the upper ship the pace of the lower ships clocks seemed to slow down as seen from the upper ship. Since the time the ships were all at rest with all clocks on both ships running at the same pace to the time on the upper ships' clocks shown in the diagram above the upper ship clocks had all advance by 270.05 ns whereas the lower ship clocks had advanced much less: 50ns, 77.06ns, and 104.1ns.

The tables illustrate the fact that under the hypotheses of Special Relativity there is reciprocal compression of lengths and reciprocal clock slowing regardless which of two bodies has received the force of acceleration which put the bodies into motion with respect to each other.

In summary, the discussion and tables above show that, according to the hypotheses of Einstein's Relativity, when in a vacuum one of two bodies which have been at rest with respect to each other with synchronized clocks is accelerated apart for a period by a force then, measurements by measuring rods and simultaneous clocks on either body will indicate that at the end of the acceleration period the length of the other body has been reduced, the reading of the clock at the center of gravity of the other body is behind the reading of the clocks on the measuring body, clocks behind the clocks at the center of gravity of the other body have readings ahead of the reading of the clock at that ship's center of gravity, clocks ahead of the clock at the center of gravity of the other ship will have readings behind the clock at the center of gravity, and the readings of the clocks on the other body will all be advancing at a common pace slower than the pace of the clocks on the measuring body. The illustration also indicates that as a result of the hypotheses of Special Relativity both bodies would measure the same speed of any light approaching them regardless of the velocity of the source of the light.

An illustration similar to that above could be prepared to demonstrate that the Relativity hypotheses also lead to the velocity of two light rays being measured at the same velocity by any target of those rays even though those rays were emitted by bodies moving with respect to each other.

The new model proposed in the first section of this document does not include all of the Relativity hypotheses described above for the reasons summarized in the notes on why this document was written.

6

Comments on Two Light Velocity Experiments

Before November 1887 most scientists believed that all light, including reflected light, moved at a constant velocity, c, with respect to an ether which was believed to be at least roughly stationary with respect to the sun. Consequently all light flashes on the earth were believed to move at an approximately constant speed with respect to the sun. A corollary of this belief was the belief that the velocity of reflected light is not influenced by the velocity of the reflecting surface. Then in the American Journal of Science of November 1887 Albert Michelson and Edward Morley reported an experiment which they had conducted using a sensitive apparatus which they had developed. That experiment had been expected to show, for example, that a light flash moving tangentially with respect to the earth and entering the apparatus in the direction opposite to the movement of the earth and the apparatus in orbit about the sun would have a higher velocity with respect to the apparatus than a light flash coming to the apparatus from other directions. But the experiment didn't work out that way. Light coming from all directions seemed to enter the apparatus at the same velocity. Belief in the ether was generally abandoned, and in 1905 most physicists adopted the hypothesis of Einstein's Special Relativity that light in a vacuum should be measured at the same velocity, c, by every observer wherever he was whatever his movement. The results of the 1887 experiment, and of more refined similar later experiments, do not, however, rule out the possibility that most light flashes on the earth do have about the same high velocity with respect to the earth, and not with respect to things moving with respect to the earth, if the velocity of reflected light is influenced by the velocity of the reflecting surface. This is the assumption adopted in the proposed model.

The behavior of light on the earth assumed in the model could have resulted from most photons on the earth having come originally to the earth from the sun at a velocity of c relative to the sun. If such a photon coming vertically toward the earth's surface were, for example, reflected to the north at noon by a mirror stationary with respect to the equator the orbital motion of the earth to the west around the sun, and the lesser spinning motion of the earth to the east, would not cause any change in that photon's velocity, c, relative to the sun, and the resulting velocity of that photon relative to the earth would also be c. If an incoming photon from the sun reached the earth at noon at the equator and were reflected eastward then that photon would be given a new velocity less than c with respect to the sun but equal to c with respect to the earth. If an incoming photon were reflected westward at noon by a mirror on the equator that photon would be given a velocity greater than c toward the west but equal to c with respect to the earth.

A light flash approaching a reflector on the earth's surface late in the afternoon and then reflected eastward in the direction the earth was spinning would have a higher reflected velocity relative to the sun than one reflected westward early in the morning.

The difference just mentioned in the velocities of the east-bound and west-bound photons relative to the sun would not be great since the orbital velocity of the earth relative to the sun is only about $1/10,000^{th}$ of c, and the spin velocity of the earth's surface is even less, about $1/548,446^{th}$ of c, but the difference would exist even if it is not often noticed on the earth.

In addition to the differences in the velocity of light relative to the sun just mentioned, Professors Petr Beckmann and Howard Hayden have pointed out evidence that all rays of light do not always have the same velocity relative to the earth. Experiments of Sagnac in 1913, of Michelson Gale in 1924, of Hafele Keating in 1971, of Brillet Hall in 1979, and of Allan, Weiss and Ashby in 1985 revealed differences between the velocities of Eastbound and Westbound Coulomb forces. The supposition of Beckmann and Hayden was that as the earth rotates Eastward the lines of gravitational attraction toward the earth do not rotate and that those lines of attraction then retard photons traveling Eastward above the earth and to some extent pull along photons traveling Westward.

The proposed model's assumption with respect to the velocity of light on the earth is consistent with the model's assumption that the velocity of reflected light is influenced by the velocity of the reflecting surface. That assumption is, however, in conflict with the view which has been conventional at least since another influential report by Michelson in the

Astrophysical Journal of April 1913 on an experiment he had conducted. On the basis of that experiment Michelson concluded "that within the limit of error of experiment (say 2 per cent) the velocity of a moving mirror is without influence on the velocity of light reflected from its surface." In the author's view Michelson's interpretation of that experiment was defective, as explained below.

In his 1913 experiment Michelson used an apparatus the essential aspects of which are shown in the following diagram.

Diagram 10

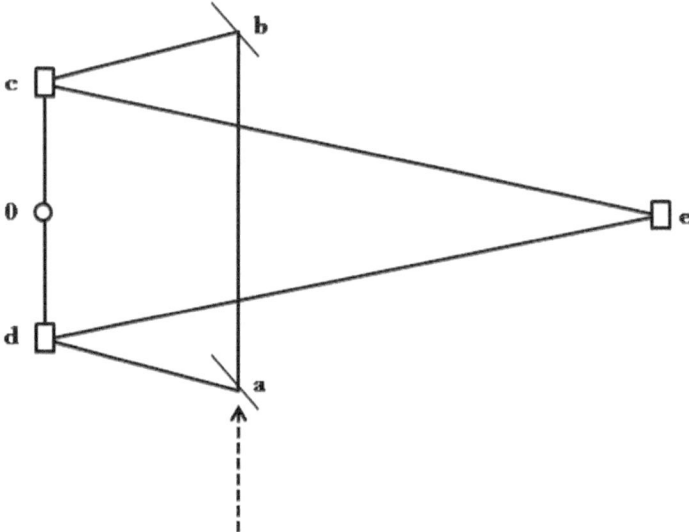

In the experiment light coming from below fell on a lightly silvered mirror at **a**. Of that light a portion was reflected to the mirror at **d**. That mirror was on a beam which was rotating counterclockwise about a point **o**, and another mirror was on the upper end of that beam. The light arriving at **d** was reflected to the mirror at **e**, then reflected to the mirror at **c**, further reflected to the mirror at **b**, and finally reflected back to point **a**. The other portion of the light originally arriving at **a** passed through the mirror at **a** and then proceeded to **b**, to **c**, to **e**, to **d**, and back to **a**, where interference fringes were produced to be observed by means of a telescope with micrometer eyepiece.

Both portions of light traveled at the same velocity when traveling between **a** and **d**, between **c** and **b**, and between **b** and **a**. But because of the movement of the mirrors on the rotating beam the light portion traveling counterclockwise from **d** to **e** to **c** traveled a longer distance than the light portion traveling clockwise from **c** to **e** to **d.** For this reason the light portions could, depending on their velocities, arrive back at point **a** at different times producing interference fringes.

The centers of the mirrors on the beam were 26.5 centimeters (cm) apart and the distance from **d** to **e** was 608.144 cm. For the experiment the beam was made to rotate 1000 revolutions per minute, giving the beam mirrors a velocity of 83,252 cm per minute. Light with a wavelength of .0006 in cm was used. With this apparatus and this light the two portions of light arrived back at **a** 3.81 wavelengths apart on average, that is with a fringe shift of 3.81.

Separately Michelson calculated what the fringe shifts should have been on two different assumptions as to how light is reflected from a moving reflecting surface. Apparently he assumed the incoming light was moving at a c of 299,212.4 km/sec, that is 179,527,440,000cm/min. He calculated that if movement of the reflecting surfaces had no effect on the velocity of the reflected light then the fringe shift should have been 3.76. If movement of the reflecting surface had a double effect on the velocity of the reflected light, as happens in the reflection of macro bodies in an elastic collision, he concluded that there would be no fringe shift at all. Since his observed fringe shift was close to his calculated fringe shift if the movement of reflecting surface had no effect he concluded that there was in fact no effect of movement of the reflecting surface on the velocity of the reflected light.

The Michelson conclusion is not persuasive to the author, however, since Michelson made significant simplifications in carrying out his calculations. His calculations did not take into account the distance the counterclockwise portion of light had to travel while the mirror c was moving in the same direction as the light nor the distance the clockwise portion of light had to travel to cover the distance which the mirror chad traveled before the clockwise light portion was reflected.

When such simplifications are not made there is no clear message in a comparison of the calculations and the observed fringe shift. The calculations without simplifications—other than ignoring small changes in the orientation of the beam mirrors as they rotated—are shown below. They produce a calculated fringe shift of 1.88 if movement of the reflecting surfaces had no effect, and a fringe shift of .36 if movement of a reflecting surface had a double effect. Neither of these calculated shifts was close to the observed shift of 3.81.

CALCULATIONS:

(In these calculations T_1 refers to the time taken for the counterclockwise light to travel from d to c; T_2 refers to the time taken for the clockwise light to travel from c to d, v refers to the velocity of the beam mirrors; W refers to wavelength: and D refers to the distance de or ec, which Michelson considered to be the same as the distance oe.)

IF MOVEMENT OF A REFLECTING SURFACE HAS NO EFFECT:

Times:

$$T_1 = (2D + v\,T_1) / c \qquad\qquad T_2 = (2D - v\,T_2) / c$$

$$T_1 c - T_1 v = 2D \qquad\qquad T_2 c + T_2 v = 2D$$

$$T_1 = 2D / (c - v) \qquad\qquad T_2 = 2D / (c + v)$$

Time Difference:

$$T_1 - T_2 = 2D / (c - v) - 2D / (c + v)$$

Distance Difference in Centimeters:

$$[2D / (c - v) - 2D / (c + v)]\,c$$

Distance Difference in Wavelengths:

$$[2D / (c - v) - 2D / (c + v)]\,c / W$$

$$[2 \times 608.144 / (179{,}444{,}200{,}000 - 83{,}252) - 2 \times 608.144 // (179{,}444{,}200{,}000 + 83{,}252)] x c\, /.0006$$

1.88

IF MOVEMENT OF A REFLECTING SURFACE HAS DOUBLE EFFECT;

Times:

$$T_1 = (2D + v\,T_1) / (c + 2v) \qquad T_2 = (2D - v\,T_2) / (c - 2v)$$

$$T_1 c + T_1 2v - T_1 v = 2D \qquad T_2 c - T_2 v + T_2 v = 2D$$

$$T_1 = 2D / (c + v) \qquad T_2 = 2D / (c - v)$$

Time Difference:

$$T_1 - T_2 = 2D / (c + v) - 2D / (c - v)$$

Distance Difference in Centimeters:

$$2D(c + 2v) / (c + v) - 2D(c - 2v) / (c - v)$$

Distance Difference in Wavelengths:

$$[2D(c + 2v) / (c + v) - 2D(c - 2v) / (c - v)] / W$$

[2 x 608.144 x (179,444,200,000 + 166,504.8)/
(179,444,200,000+83,252.4) - 2 x 608 x (179,444,200,000 -
166,504.8) / (179,444,200,000-83,252.4)] /.0006 = .36

www.ingramcontent.com/pod-product-compliance
Lightning Source LLC
Chambersburg PA
CBHW022011170526
45157CB00003B/1224